T0348649

Methods in Cell Biology

Micropatterning in Cell Biology
Part C

Volume 121

Series Editors

Leslie Wilson
Department of Molecular, Cellular and Developmental Biology
University of California
Santa Barbara, California

Phong Tran
Department of Cell and Developmental Biology
University of Pennsylvania
Philadelphia, Pennsylvania

Methods in Cell Biology

Micropatterning in Cell Biology
Part C

Volume 121

Edited by

Matthieu Piel

Systems Cell Biology of Cell Division and Cell Polarity, UMR144, Institut Curie, CNRS, Paris, France

Manuel Théry

Laboratoire de Physiologie Cellulaire et Végétale, Institut de Recherches en Technologies et Sciences, pour le Vivant, iRTSV, CNRS/CEA/INRA/UJF, Grenoble, France

AMSTERDAM • BOSTON • HEIDELBERG • LONDON
NEW YORK • OXFORD • PARIS • SAN DIEGO
SAN FRANCISCO • SINGAPORE • SYDNEY • TOKYO
Academic Press is an imprint of Elsevier

Academic Press is an imprint of Elsevier
525 B Street, Suite 1800, San Diego, CA 92101-4495, USA
225 Wyman Street, Waltham, MA 02451, USA
The Boulevard, Langford Lane, Kidlington, Oxford, OX5 1GB, UK
32 Jamestown Road, London NW1 7BY, UK
Radarweg 29, PO Box 211, 1000 AE Amsterdam, The Netherlands

First edition 2014

ISBN: 978-0-12-800281-0
ISSN: 0091-679X

For information on all Academic Press publications visit
our website at store.elsevier.com

Printed and bound by CPI Group (UK) Ltd, Croydon, CR0 4YY

Working together
to grow libraries in
developing countries

www.elsevier.com • www.bookaid.org

Contents

v

SECTION 2   CELL AND TISSUE MICROPATTERNING IN 3D

Contributors

Saniya Ali
Department of Biomedical Engineering, Duke University, Durham, North Carolina, USA

Ryan C. Bailey
Departments of Chemistry, University of Illinois at Urbana-Champaign, Urbana, Illinois, USA

Daria Bonazzi
Systems Cell Biology of Cell Division and Cell Polarity, Institut Curie, CNRS, Paris, France

Thomas Boudou
Laboratory of Materials and Physical Engineering, CNRS UMR5628, and Grenoble Institute of Technology, Grenoble, France

Christopher S. Chen
Department of Bioengineering, University of Pennsylvania, Philadelphia, Pennsylvania, USA

Steffen Cosson
Laboratory of Stem Cell Bioengineering, Institute of Bioengineering, School of Life Sciences, Ecole Polytechnique Fédérale de Lausanne, Lausanne, Switzerland

Maude L. Cuchiara
Department of Biomedical Engineering, Duke University, Durham, North Carolina, USA

Jean Marc Di Meglio
Laboratoire Matière et Systèmes Complexes, UMR7057, CNRS & Université Paris Diderot, Paris, France

Fanny Evenou
Laboratoire Matière et Systèmes Complexes, UMR7057, CNRS & Université Paris Diderot, Paris, France

Shirin Feghhi
Department of Mechanical Engineering, University of Washington, Seattle, Washington, USA

Sylvain Gabriele
Mechanobiology and Soft Matter Group, Laboratoire Interfaces et Fluides Complexes, Complexys and Biosciences Research Institutes, CIRMAP, Université de Mons, Mons, Belgium

David H. Gracias
Department of Chemical and Biomolecular Engineering, and Department of Chemistry, Johns Hopkins University, Baltimore, Maryland, USA

Thomas Grevesse
Mechanobiology and Soft Matter Group, Laboratoire Interfaces et Fluides
Complexes, Complexys and Biosciences Research Institutes, CIRMAP,
Université de Mons, Mons, Belgium

Wei-hui Guo
Department of Biomedical Engineering, Carnegie Mellon University, Pittsburgh,
Pennsylvania, USA

Sangyoon J. Han
Department of Mechanical Engineering, University of Washington, Seattle,
Washington, USA

Brendan A. Harley
Chemical and Biomolecular Engineering, University of Illinois at
Urbana-Champaign, Urbana, Illinois, USA

Pascal Hersen
Laboratoire Matière et Systèmes Complexes, UMR7057, CNRS & Université Paris
Diderot, Paris, France

Ian Hoffecker
Department of Biomedical Engineering, Carnegie Mellon University, Pittsburgh,
Pennsylvania, USA

Daniel Irimia
Massachusetts General Hospital, Harvard Medical School, and Shirners Hospitals
for Children, Boston, Massachusetts, USA

Sahar Javaherian
Department of Chemical Engineering and Applied Chemistry, University of
Toronto, Toronto, ON, Canada

L. Kőhidai
Department of Genetics, Cell- and Immunobiology, Semmelweis University,
Budapest, Hungary

C.H.R. Kuo
Biological and Soft Systems, Cavendish Laboratory, University of Cambridge,
Cambridge, United Kingdom

Joséphine Lantoine
Mechanobiology and Soft Matter Group, Laboratoire Interfaces et Fluides
Complexes, Complexys and Biosciences Research Institutes, CIRMAP,
Université de Mons, Mons, Belgium

Franziska Lautenschlaeger
Systems Cell Biology of Cell Division and Cell Polarity, Institut Curie, CNRS, Paris,
France

Maël Le Berre
Systems Cell Biology of Cell Division and Cell Polarity, Institut Curie, CNRS, Paris,
France

Wesley R. Legant
HHMI Janelia Farm Research Campus, Ashburn, Virginia, USA

J. Láng
Biological and Soft Systems, Cavendish Laboratory, University of Cambridge, Cambridge, United Kingdom, and Department of Genetics, Cell- and Immunobiology, Semmelweis University, Budapest, Hungary

O. Láng
Department of Genetics, Cell- and Immunobiology, Semmelweis University, Budapest, Hungary, and Institute for Integrated Cell-Material Sciences, Kyoto University, Kyoto, Japan

Matthias P. Lutolf
Laboratory of Stem Cell Bioengineering, Institute of Bioengineering, School of Life Sciences, Ecole Polytechnique Fédérale de Lausanne, Lausanne, Switzerland

Alison P. McGuigan
Department of Chemical Engineering and Applied Chemistry, and Institute for Biomaterials and Biomedical Engineering, University of Toronto, Toronto, ON, Canada

Ana C. Paz
Department of Chemical Engineering and Applied Chemistry, and Institute for Biomaterials and Biomedical Engineering, University of Toronto, Toronto, ON, Canada

Catherine Picart
Laboratory of Materials and Physical Engineering, CNRS UMR5628, and Grenoble Institute of Technology, Grenoble, France

Matthieu Piel
Systems Cell Biology of Cell Division and Cell Polarity, Institut Curie, CNRS, Paris, France

Samuel R. Polio
Department of Biomedical Engineering, Boston University, Boston, Massachusetts, USA

Alexandre Ramade
Laboratory of Materials and Physical Engineering, CNRS UMR5628, and Grenoble Institute of Technology, Grenoble, France

Alexander Revzin
Department of Biomedical Engineering, University of California, Davis, California, USA

Maryam Riaz
Mechanobiology and Soft Matter Group, Laboratoire Interfaces et Fluides Complexes, Complexys and Biosciences Research Institutes, CIRMAP, Université de Mons, Mons, Belgium

Dong-Sik Shin
Department of Biomedical Engineering, University of California, Davis, California, USA

E. Sivaniah
Biological and Soft Systems, Cavendish Laboratory, University of Cambridge, Cambridge, United Kingdom, and Institute for Integrated Cell-Material Sciences, Kyoto University, Kyoto, Japan

Michael L. Smith
Department of Biomedical Engineering, Boston University, Boston, Massachusetts, USA

Nathan J. Sniadecki
Department of Mechanical Engineering, and Department of Bioengineering, University of Washington, Seattle, Washington, USA

Lucas H. Ting
Department of Mechanical Engineering, University of Washington, Seattle, Washington, USA

Aurora J. Turgeon
Departments of Chemistry, University of Illinois at Urbana-Champaign, Urbana, Illinois, USA

Marie Versaevel
Mechanobiology and Soft Matter Group, Laboratoire Interfaces et Fluides Complexes, Complexys and Biosciences Research Institutes, CIRMAP, Université de Mons, Mons, Belgium

Clément Vulin
Laboratoire Matière et Systèmes Complexes, UMR7057, CNRS & Université Paris Diderot, Paris, France

Yu-li Wang
Department of Biomedical Engineering, Carnegie Mellon University, Pittsburgh, Pennsylvania, USA

Jennifer L. West
Department of Biomedical Engineering, Duke University, Durham, North Carolina, USA

Stephanie Wong
Department of Biomedical Engineering, Carnegie Mellon University, Pittsburgh, Pennsylvania, USA

Zinnia S. Xu
Department of Biomedical Engineering, Johns Hopkins University, Baltimore, Maryland, USA

Cem Onat Yilmaz
Department of Materials Science and Engineering, Johns Hopkins University, Baltimore, Maryland, USA

Jungmok You
Department of Biomedical Engineering, University of California, Davis, California, USA

Ewa Zlotek-Zlotkiewicz
Systems Cell Biology of Cell Division and Cell Polarity, Institut Curie, CNRS, Paris, France

Preface

Micropatterning refers generally to techniques which provide an experimental control over the chemical, physical, or geometrical properties of materials at the micron or submicron scale, and are thus used to produce spatial patterns of these properties. These techniques, which were often originally designed for application in microelectronics, have spread over most areas of science, including biology. They have proved particularly useful for cell biology, bridging the gap between the Petri-dish and complex 3D assays and tissues. At the level of single cells, many environmental parameters are entangled and assessing their individual contribution to cell physiology and behavior is often difficult. Micropatterned cell-culture substrates allow to specifically design tools to quantitatively control the cell microenvironment *in vitro* and to assess the effect of individual parameters, with devices which are almost as easy to handle as a regular Petri dish. Historically, printing of cell adhesion molecules, such as collagen or fibronectin, have allowed producing cell culture substrates on which cells have a well-defined shape and adhesion geometry. Such substrates have been crucial to demonstrate the role of cell shape, cell spreading area and of geometrical parameters of cell adhesion on cell survival, proliferation, differentiation, and polarity. The fabrication of micropatterned substrates initially required special expertise in surface chemistry and sophisticated devices, but their success lead to the development of much simpler methods accessible to almost any regular biology lab (see Chapters 1–6 of vol. 119 for printing of proteins on various types of substrates, including printing of multiple proteins and of gradients). Efforts have also been made to make the process cheaper and more versatile (see maskless techniques in Chapters 7–11 of vol. 119). Micropatterning now covers a large number of cell biology applications, from stem cell culture and differentiation (see e.g., Chapters 2 and 13 of vol. 119) to printing of purified proteins or other biomolecules for *in vitro* assays (see Chapters 15 of vol. 119 and 1–4 of vol. 120). Moreover, the size of the features which can be printed is now down to tens of nanometers (see Chapters 12–14 of vol. 119). Current micropatterning techniques have developed further to implement the quantitative control of other aspects of the cell microenvironment such as 3D geometry (see Chapters 7–15) and mechanical properties (see Chapters 16 of vol. 120, 3, and 6). Importantly, some of these tools do not only allow building microcontrolled environments for cultured cells, but are also measurement tools, giving access to crucial parameters such as forces (see Chapters 13 of vol. 120, 1, 2, and 4). Although the technical basis for most micropatterning methods is very generic, clever variations and adaptation are enough to produce tools for very specific applications, such as the study of collective cell behavior (see Chapter 15 of vol. 120), imaging of yeast cells from the tip (Chapter 14 of vol. 120), or local application of forces on individual cells

(Chapter 12 of vol. 120). The latest evolutions of micropatterning are meant to implement temporal control of the micropatterned features (see Chapters 5–11 of vol. 120), to reach full spatio-temporal control of the cell microenvironment.

Matthieu Piel and Manuel Théry

Micropatterning Soft Substrates

Preparation of a Micropatterned Rigid-Soft Composite Substrate for Probing Cellular Rigidity Sensing

1

Stephanie Wong, Wei-hui Guo, Ian Hoffecker, and Yu-li Wang

Department of Biomedical Engineering, Carnegie Mellon University, Pittsburgh,
Pennsylvania, USA

CHAPTER OUTLINE

Abstract

Substrate rigidity has been recognized as an important property that affects cellular physiology and functions. While the phenomenon has been well recognized, understanding the underlying mechanism may be greatly facilitated by creating a microenvironment with designed rigidity patterns. This chapter describes in detail an optimized method for preparing substrates with micropatterned rigidity, taking advantage of the ability to dehydrate polyacrylamide gels for micropatterning with photolithography, and subsequently rehydrate the gel to regain the original elastic state. While a wide range of micropatterns may be prepared, typical composite substrates consist of micron-sized islands of rigid photoresist grafted on the surface of polyacrylamide hydrogels of defined rigidity. These islands are displaced by cellular traction forces, for a distance determined by the size of the island, the rigidity of the underlying hydrogel, and the magnitude of traction forces. Domains of rigidity may be created using this composite material to allow systematic investigations of rigidity sensing and durotaxis.

INTRODUCTION AND RATIONALE

Micropatterning has been utilized during the past two decades to create microenvironments of defined geometry at a micron scale (Whitesides, Ostuni, Takayama, Jiang, & Ingber, 2001). It allows systematic testing of specific features of the *in vivo* environment for their biological effects. By controlling the geometry of adhesion areas on glass, previous micropatterning studies have demonstrated the effects of cell shape and size on events such as apoptosis, proliferation, differentiation, and migration (Chen, 1997; Dike et al., 1999; McBeath, Pirone, Nelson, Bhadriraju, & Chen, 2004; Pouthas et al., 2008; Singhvi et al., 1994; Wang, Ostuni, Whitesides, & Ingber, 2002).

An important parameter that would benefit from micropatterning studies is substrate rigidity, which is known to cause profound cellular responses (Discher, Janmey, & Wang, 2005; Engler, Sen, Sweeney, & Discher, 2006; Pelham & Wang, 1997). Most studies of cellular rigidity sensing have relied on the use of either elastic polymers or bendable micropost arrays as the substrate. The former included polydimethylsiloxane (PDMS) and polyacrylamide, where the elasticity may be controlled over a wide range by altering the concentration of the crosslinker and/or the base material. The latter involved the preparation of PDMS pillars with bending moduli varied over a limited range by changing the diameter or height of the pillars (Tan et al., 2002; Trichet et al., 2012), or over a wider range by changing both the rigidity of PDMS used for pillar fabrication and dimension of the pillars (Sun, Jiang, Okada, & Fu, 2012).

Both elastic polymers and micropost arrays may be micropatterned by selective coating of the surface with adhesive proteins to control cell size, shape, and migration. Methods used include microstencils (Wang et al., 2002), microcontact printing (Théry & Piel, 2009), activation with deep UV exposure though a mask (Tseng et al., 2011), and microcontact printing of activated proteins on glass followed by transfer of the pattern to the elastic substrate (Rape, Guo, & Wang, 2011). What has been lacking, however, is a method to create micropatterns of mixed rigidity, given the importance of studying cellular responses at a rigidity interface such as in durotaxis (Lo, Wang, Dembo, & Wang, 2000). Previous methods to address durotaxis have involved the creation of a border of rigidity across the substrate surface (Trichet et al., 2012; Wang, Dembo, Hanks, & Wang, 2001), with serious limitations in the number cells that may be studied at the border.

Several methods have been developed for creating a rigidity interface across the substrate surface, including polymerization of hydrogels using photosensitive reactions and patterned UV illumination (Nemir, Hayenga, & West, 2010; Wong, Velasco, Rajagopalan, & Pham, 2003), and overlay of a thin hydrogel layer on micron-sized rigid topographic features (Choi et al., 2012; Gray, Tien, & Chen, 2003). Rigidity changes may also be created locally in real time using hydrogels formed with photo-labile crosslinkers (Frey & Wang, 2009; Kloxin, Benton, & Anseth, 2010). While these methods create fixed rigidity domains of similar adhesiveness, different questions may be addressed by creating small mobile islands of rigid adhesive materials grafted onto soft nonadhesive surfaces. In the initial study using this approach, we showed that long-range substrate strain between the islands dominates over local rigidity, in determining cellular responses (Hoffecker, Guo, & Wang, 2011).

To prepare such composite substrates, we take advantage of the ability to dehydrate polyacrylamide hydrogels, for the attachment of photoresist and micropatterning using photolithography, and subsequently rehydrate the hydrogel to regain the original elastic property. The polyacrylamide surface remains nonadhesive to cells, while the photoresist may be treated with extracellular matrix proteins to enhance cell adhesion. We present here a detailed method for generating this material, using only inexpensive equipment without a clean room.

1.1 MATERIALS

1. Coverslip (45 mm × 50 mm #1; Fisher Scientific, Pittsburgh, PA)
2. Diamond tip pen
3. Bunsen burner
4. Bind silane working solution: Mix 950 mL, 95% ethanol, and 50 mL of 95% glacial acetic acid, and add 3 μL of bind silane (γ-methacryloxypropyltrimethoxysilane; GE Healthcare, Waukesha, WI) to form the working solution
5. Ethanol (C_2H_5OH) 95%
6. Acrylamide solution (40% w/v; BioRad, Hercules, CA)
7. Bis-acrylamide solution (2% w/v; BioRad)
8. N,N,N',N'-tetramethylethylenediamine (TEMED; BioRad)
9. Ammonium persulfate aqueous solution (APS, 10% w/v; BioRad)
10. $10\times$ and $1\times$ phosphate buffered saline (PBS, pH 7.4)
11. Aqueous sucrose solution (50% w/v)
12. Rain-X
13. Coverslip (#2, 25 mm circular or square; Fisher Scientific)
14. Razor blade
15. Tweezers
16. Heating block set to 95 °C
17. SU-8 2002 (MicroChem Corp., Newton, MA)

18. SU-8 developer (MicroChem Corp.)
19. Spin coater
20. UV light source for crosslinking SU-8
21. Orbital shaker
22. Photomasks, on inexpensive transparency films for patterns larger than 10 μm or chrome plated lime glass for smaller features
23. Glass Petri dishes

1.2 METHODS

Our method involves the following steps: (A) Preparation of thin sheets of polyacrylamide of defined rigidity covalently bound to a glass coverslip for stability (Fig. 1.1A); (B) dehydration of the polyacrylamide sheet to allow grafting and

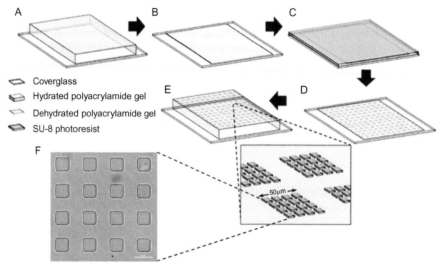

FIGURE 1.1

Schematic depicting the steps involved in composite substrate fabrication.
(A) Polyacrylamide hydrogels are polymerized on bind-silane-activated glass coverslips.
(B) Hydrogels are air dried to dehydrate and flatten the hydrogel before micropatterning.
(C) A thin layer of SU-8 is spin coated on the dehydrated hydrogel surface. (D) The surface is patterned via UV exposure through a photomask containing the desired pattern and developed to remove unexposed regions of SU-8 leaving behind the desired pattern.
(E) Rehydration of the polyacrylamide in PBS allows the hydrogel to reswell with the patterned SU-8 islands grafted to the surface. (F) Phase contrast image of the composite substrate consisting of a 4 × 4 array of small SU-8 islands on a polyacrylamide gel surface. (See color plate.)

Reproduced from Hoffecker et al. (2011).

micropatterning of SU-8 using photolithography (Fig. 1.1B–D); (C) rehydration of the polyacrylamide sheet and coating of the surface of SU-8 to promote cell adhesion (Fig. 1.1E). The pattern of SU-8 grafted to the polyacrylamide gel may be easily seen using transmitted optics (Fig. 1.1F).

1.2.1 Preparation of polyacrylamide gel base

A thin sheet of polyacrylamide gel serves as an elastic base for anchoring rigid islands of SU-8 photoresist, such that translocation of the islands is determined by the rigidity of polyacrylamide, the size of the islands, and forces applied by cells to the adhesive islands. The sheet of polyacrylamide must be covalently bound to a glass surface during dehydration, when the hydrogel sheet would otherwise shrink and detach from the surface. Bonding is established by pretreating the glass surface with bind silane (γ-methacryloxypropyltrimethoxysilane), which reacts with glass through the trimethoxysilane moiety and copolymerizes with acrylamide through the methacryl moiety. For this procedure, one side of a coverslip (45 mm × 50 mm in our applications) is marked with a diamond tip pen for identification, the marked side is passed through the plasma of a Bunsen burner flame to render the surface hydrophilic, and 30 μL of bind silane working solution is spread evenly over the flamed side of the coverslip, in a fume hood, using a cotton swab. After 15 min at room temperature, when the surface has fully dried, the treated side is rinsed with 70% ethanol and allowed to air dry. These activated coverslips can be stored at room temperature in a desiccator for at least 3 months.

A solution of acrylamide and bis-acrylamide is then polymerized on the activated glass surface. A uniform sheet is formed by polymerizing a small volume underneath a top coverslip (25 mm circular or square in our applications). The use of a top coverslip also restricts oxygen exposure, which inhibits the polymerization of acrylamide. A very thin sheet, formed by using a volume of 0.03–0.04 μL/mm^2 surface area during polymerization, is desirable for compatibility with microscope optics. In addition, a thin gel barely rises above the glass surface after dehydration, which facilitates spin coating and tight contact with the photomask during photolithography (Section 1.2.2).

Removal of the top coverslip is facilitated by coating its surface with either Rain-X (to increase hydrophobicity) or 50% sucrose solution (to serve as a separation layer). Such treatment is essential to minimize the adhesion of the polyacrylamide surface to the top coverslip, which would otherwise introduce defects on the gel surface, particularly for soft gels, during the removal of the top coverslip. Surface defects in turn compromise the quality of micropatterning and interfere with microscopy. Rain-X solution is smeared over the coverslip and wiped away with a Kimwipe. The surface is then rinsed with distilled water and wiped until clear. It is important to ensure complete treatment of the surface with Rain-X to prevent any local adhesion of the coverslip to polyacrylamide.

To coat the surface with sucrose solution, one side of a top coverslip is marked with a permanent marker and the unmarked side is passed through the plasma of a Bunsen burner flame to render the surface hydrophilic and ensure an even coating of sucrose solution. Approximately 100 μL of 50% w/v sucrose solution is then placed on the flamed side of the coverslip and spread over the entire surface. Uniform coating is achieved by placing the coverslip in a spin coater and spinning the sample at 5000 rpm for 15 s.

The rigidity of the polyacrylamide sheet is controlled by changing the concentration of acrylamide and/or bis-acrylamide (Pelham & Wang, 1997). Typically, calculated volumes of acrylamide, bis-acrylamide, and 0.1 mL of $10\times$ PBS are mixed with distilled water to reach a total volume of 1 mL and then placed in a sealed chamber under house vacuum to degas for 30 min. Skin contact with unpolymerized acrylamide should be avoided as it is a neurotoxin. Freshly prepared 10% w/v APS and TEMED are then added at a volume of 6 and 4 μL, respectively, and mixed by gentle tapping or pipetting. A 20 μL droplet of solution is pipetted immediately onto a bind-silane-activated coverslip and a Rain-X or sucrose-coated top coverslip is placed on top using a pair of fine tweezers, with the coated side facing the acrylamide solution. The acrylamide solution should spread uniformly underneath the top coverslip; if not, gently moving the coverslip with a pair of tweezers should help. When using a sucrose-coated top coverslip, the coverslip assembly should be turned upside down during polymerization to avoid the settling of sucrose into the acrylamide solution due to its higher density. The solution is then allowed to polymerize for at least 30 min at 25 °C.

Following polymerization, the sandwich is turned right side up and the top coverslip removed carefully. Rain-X treated coverslips are removed by flooding the surface of the coverslip with distilled water and waiting for at least 15 min to allow water to seep in. A razor blade is then used to lift the top coverslip very slowly off the polyacrylamide gel, with the gel staying submerged in water, to prevent the gel from cracking due to the strain. Sucrose-coated coverslips are easier to remove and are recommended for soft gels to avoid surface cracking. They may be removed by immersing the sandwich in hot distilled water in a Petri dish to dissolve the sucrose. The coverslip should release in 20–30 min. Following the removal of the top coverslip, the gel should be equilibrated with distilled water in a Petri dish for 30 min on a shaker. This helps prevent the formation of crystals during the subsequent drying.

The edges of the polyacrylamide gel often show a slight lip that could disrupt proper micropatterning. These may be removed using a razor blade to cut away ~1 mm along the edge while keeping the gel hydrated under distilled water. The gel is then rinsed with water to remove any bits of polyacrylamide gel and allowed to air dry overnight. Complete drying of the gel is essential for the subsequent coating with SU-8 in organic solvent and photolithography.

1.2.2 Micropatterning of the polyacrylamide surface with SU-8 photoresist

Once the polyacrylamide sheet is dry, the surface is micropatterned using SU-8, a negative photoresist that polymerizes upon UV exposure (Fig. 1.1). Photomasks with a desired pattern may be obtained on either plastic transparencies (for patterns larger than 10 μm) or chrome plated lime glass (for high resolution patterns), from companies such as CAD/Art services (Bandon, OR) or Photo-Sciences (Torrance, CA), respectively, which accept CAD file formats such as dwt. Note that areas for SU-8 coverage should be clear, while areas in between should be masked.

Micropatterning is performed following standard photolithography procedures. The coverslip is first baked on a temperature-regulated heating plate for 1 min at 95 °C, to ensure that the gel surface is completely dry. After cooling to room temperature, approximately 300 μL of SU-8 photoresist solution is spread on the coverslip to cover the dehydrated gel surface. The coverslip is placed in a spin coater and spun at 5000 rpm for 20 s, then baked for 2–3 min at 95 °C and cooled to room temperature. Using SU-8 2002, this procedure should create a uniform layer ~2 μm in thickness.

Contact exposure represents the most economical way to transfer the pattern from the photomask onto the SU-8 photoresist. The coverslip is placed on a platform stand underneath a UV light source with the SU-8 side facing up, overlaid with the photomask with the patterned side facing down, and covered with a piece of plate glass 3 mm in thickness that has an area matching that of the platform. Several binder clips are placed around the plate glass to clamp the plate glass to the stand to ensure tight contact between the photomask and the coverslip. The source of UV may range from an arc lamp with collimated optics for uniform exposure (OAI, San Jose, CA) to inexpensive 360-nm UV photodiodes (Jelight, Irvine, CA). The latter should be used in conjunction with an orbital shaker, which rotates the coverslip assembly underneath the light beam to achieve uniform exposure. The exposure time is dependent on both the power of the UV source and the distance between the lamp and the coverslip, and must be determined by trial and error (see Section 1.2.4, point 2). At a light power of 100 nJ/cm^2 at the sample, the optimal exposure time should be around 60–90 s.

After exposure, the coverslip is baked for 2–3 min at 95 °C and allowed to cool to room temperature. It is then immersed in the SU-8 developer in a Petri dish for 60–90 s with agitation, rinsed briefly with ethanol from a squirt bottle, and immersed in a separate Petri dish of ethanol for ~30 s. After air drying, the micropattern should be easily visible under a microscope. There should be no speckles or films between the intended areas of SU-8, which indicate residual SU-8 due to incomplete development. In addition to potential problems with imaging, residual SU-8 may cause cell adhesion to an otherwise nonadhesive polyacrylamide surface. It may be removed by additional treatment with SU-8 developer, for approximately 10 s, and ethanol rinse as described earlier.

It is essential that all traces of SU-8 developer be removed from the hydrogel to mitigate the risk of cytotoxicity. For immediate use, the coverslip should be washed with PBS in a Petri dish for an hour with shaking, to allow both rehydration of the gel and removal of residual developer. Alternatively, the coverslips may be baked for 4 h at 95 °C to evaporate the residual developer and then stored in a dessicator at room temperature for up to 3 weeks. Repeated cycles of gel hydration and dehydration should be avoided as they can introduce microcracks on the gel surface.

1.2.3 Surface coating with extracellular matrix proteins and cell seeding

The gel of the composite substrate is rehydrated in PBS for approximately 1 h at room temperature and then sterilized under the UV of a biosafety cabinet for 15–20 min. SU-8 is known to have a biofouling surface (Voskerician et al., 2003), which passively adsorbs proteins from serum containing media. Thus, cells may attach to the SU-8 surface to some extent without treatment. Cell adhesion to SU-8 surfaces may be optimized by incubation with an extracellular matrix protein such as fibronectin, while the polyacrylamide surfaces should remain nonadhesive regardless of the incubation. A 20-min incubation at room temperature with the protein of interest (e.g. 10 μg/mL fibronectin in PBS) is usually sufficient for the promotion of cell adhesion.

Before plating cells, the composite substrate is equilibrated with culture media for at least 30 min in a CO_2 incubator. Once plated, cells should attach to areas occupied by SU-8 within 15 min, although a longer period may be required if the adhesive areas are small.

While the SU-8 pattern and adhered cells are easily visible at a low magnification with phase contrast or bright field optics, the use of high magnification oil immersion lens may be limited by the combined thickness of the hydrogel and photoresist. A lens of long working distance is essential, and better resolution may be achieved with a water immersion lens to avoid spherical aberration. It is also noteworthy that SU-8 emits autofluorescence when excited at 488 nm, therefore fluorophores with a long wavelength are preferred over probes such as Green Fluorescent Protein.

1.2.4 Troubleshooting

Although the present method may seem straightforward, problems may arise from the improper execution of a few crucial steps.

1. *Poor micropatterning*: After development, the micropattern may not look as expected. Poor contact between the photomask and SU-8 creates ill-defined borders of the micropattern. Debris on a dirty photomask, or beading of the edge of polyacrylamide or SU-8 (from poor spin coating) can create space between the photomask and SU-8. In addition, overexposure can cause SU-8 areas to appear larger than expected with poorly defined edges, while overdeveloping would cause SU-8 areas to appear smaller with rounded corners.

2. *Poor association of SU-8 with the polyacrylamide surface*: During development, the micropattern may become detached from the polyacrylamide surface. SU-8 is most likely underexposed or insufficiently heated after exposure. Underexposure prevents proper formation of crosslinks within SU-8 all the way down to the polyacrylamide surface, which grafts the photoresist to the gel surface and provides the resistance against the developer. Insufficient baking after UV exposure also prevents exposed regions of SU-8 from curing properly, while too

thick a layer of polyacrylamide may slow down heat conduction and require a longer period of baking.

3. *Adhesion of cells to supposedly nonadhesive areas of polyacrylamide*: When cells are plated they may attach in between SU-8 islands to the hydrogel surface. Most likely SU-8 is underdeveloped, which causes a thin film of unexposed SU-8 to remain on the surface of polyacrylamide. This residual film then adsorbs proteins and mediates cell adhesion. The problem may be rectified by additional treatment with SU-8 developer as noted in Section 1.2.2.

4. *Cracking of the gel beneath SU-8*: After rehydration there are cracks easily visible in the hydrogel surface between SU-8 islands. The most likely cause is adhesion of the top coverslip to the gel during removal, possibly due to incomplete coverage of the surface of top coverslip with Rain-X or sucrose solution. In addition to ensuring complete coverage, increasing the concentration of bis-acrylamide and decreasing the concentration of acrylamide may alleviate this problem by increasing gel stability while maintaining the same elastic modulus.

5. *Poor cell adhesion*: During plating the cells may not quickly adhere to the SU-8 islands. If plated cells fail to attach to SU-8 within 20–30 min, a likely cause is insufficient time of incubation with extracellular matrix proteins or serum containing medium.

1.3 DISCUSSION

A significant advantage of the procedure above is the use of inexpensive equipment, including high flux UV-LED as the light source in conjunction with an orbital shaker for uniform illumination (Guo & Wang, 2011). In addition, as pointed out by Tsai, Crosby, and Russell (2007), most biological applications can tolerate some defects in the micropattern, which allows photolithography to be performed without a clean room facility.

The effects of the above procedures on rheological properties of the polyacrylamide hydrogel have been assessed with sheets ~500 µm in thickness using a Bohlin Gemini Advanced Rheometer (Malvern Instruments Inc., NJ). No significant difference in shear modulus was found between untreated hydrogels and hydrogels subjected to the micropatterning procedure without UV exposure (thus removal of the entire layer of SU-8 during the development), suggesting that the process of micropatterning, including dehydration and rehydration, does not affect mechanical properties of the hydrogel.

To determine if cellular traction forces may cause small SU-8 islands to dislodge from the surface of polyacrylamide hydrogels, cells were allowed to adhere overnight to a regular array of islands on a soft hydrogel. While traction forces caused visible distortion of the pattern, the islands returned to their regular positions upon removal of the cell with trypsin (Fig. 1.2). This indicates that the SU-8 islands are well adhered to the polyacrylamide hydrogel, and that cellular traction forces do not cause slippage between the gel and islands.

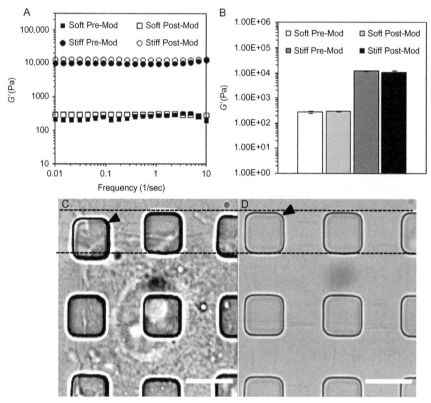

FIGURE 1.2

(A) The shear modulus of a stiff and soft hydrogel, before and after the micropatterning procedure without UV exposure, was measured using parallel plate rheometry at a constant strain of 0.1 as a function of frequency. As the frequency changes, the mechanical properties of the gel are not significantly changed before or after modification. (B) Average values for shear modulus of a stiff and soft gel show that there is no significant change in gel mechanical properties from the micropatterning step. Error bars represent standard error of the mean (SEM). (C) To assess the elastic recovery of the hydrogel, cells were cultured overnight on the patterned substrate. Displacement of the island is evident, yellow lines and arrow, from the force exerted on the island by the cell. (D) After removal of the cell with trypsin the island recovers to the original position evidenced by yellow line and arrow. This illustrates the elastic recovery of the gel and that there is no slippage of the island on the gel when force is exerted on the island. (For interpretation of the references to color in this figure legend, the reader is referred to the online version of this chapter.)

Reproduced from Hoffecker et al. (2011).

Compared to homogeneous substrates, composite materials prepared with the present procedure are particularly advantageous for studying cellular responses to rigidity. Both the stiffness of the hydrogel base and the pattern of adhesive photoresist can be varied according to the experimental design. For example, the spatial resolution for cellular rigidity sensing has been assessed by varying the size of the island

and the distance between islands. The results demonstrated that long-range strains dominate cellular responses over local rigidity (Hoffecker et al., 2011; Trichet et al., 2012), although other approaches suggested that local rigidity may also play a role under some conditions (Ghassemi et al., 2012; Sun et al., 2012). Future applications of composite substrates may help dissect the mechanism of rigidity sensing at both spatial regimes.

Recent studies have demonstrated the importance of mechanical microenvironments in regulating cell physiology, differentiation, migration, and organization (Discher et al., 2005; Engler et al., 2006). Micropatterning methods are emerging as an important tool for engineering microenvironments *in vitro* to mimic the conditions *in vivo* or dissect the effects of specific parameters. The present method may be modified for different applications. For example, while cell adhesion is limited to the SU-8 surface following the present procedure, the entire surface including polyacrylamide may be rendered adhesive using activation reagents such as Sulfo-SANPAH (Pierce, cat. No. 22589, Rockford, IL). Cells would then probe the substrate continuously, rather than respond to the deformability between rigid SU-8 domains.

Acknowledgment

Development of the present method was supported by a grant GM-32476 from NIH to YLW.

References

Chen, C. S. (1997). Geometric control of cell life and death. *Science, 276,* 1425–1428.

Choi, Y. S., Vincent, L. G., Lee, A. R., Kretchmer, K. C., Chirasatitsin, S., Dobke, M. K., et al. (2012). The alignment and fusion assembly of adipose-derived stem cells on mechanically patterned matrices. *Biomaterials, 33,* 6943–6951.

Dike, L. E., Chen, C. S., Mrksich, M., Tien, J., Whitesides, G. M., & Ingber, D. E. (1999). Geometric control of switching between growth, apoptosis, and differentiation during angiogenesis using micropatterned substrates. *In Vitro Cellular and Developmental Biology—Animal, 35,* 441–448.

Discher, D. E., Janmey, P., & Wang, Y.-L. (2005). Tissue cells feel and respond to the stiffness of their substrate. *Science (New York, NY), 310,* 1139–1143.

Engler, A. J., Sen, S., Sweeney, H. L., & Discher, D. E. (2006). Matrix elasticity directs stem cell lineage specification. *Cell, 126,* 677–689.

Frey, M. T., & Wang, Y.-L. (2009). A photo-modulatable material for probing cellular responses to substrate rigidity. *Soft Matter, 5,* 1918–1924.

Ghassemi, S., Meacci, G., Liu, S., Gondarenko, A. A., Mathur, A., Roca-Cusachs, P., et al. (2012). Cells test substrate rigidity by local contractions on submicrometer pillars. *Proceedings of the National Academy of Sciences, 109,* 5328–5333.

Gray, D. S., Tien, J., & Chen, C. S. (2003). Repositioning of cells by mechanotaxis on surfaces with micropatterned Young's modulus. *Journal of Biomedical Materials Research, 66A,* 605–614.

Guo, W., & Wang, Y.-L. (2010). Micropatterning cell-substrate adhesions using linear poly-acrylamide as the blocking agent. In R.D. Goldman, J.R. Swedlow, & D.L. Spector (Eds.), *Live Cell Imaging* (2nd ed., pp. 43–52). Cold Spring Harbor, NY: CHSP.

Hoffecker, I. T., Guo, W., & Wang, Y. (2011). Assessing the spatial resolution of cellular rigidity sensing using a micropatterned hydrogel-photoresist composite. *Lab on a Chip, 11,* 3538–3544.

Kloxin, A. M., Benton, J. A., & Anseth, K. S. (2010). In situ elasticity modulation with dynamic substrates to direct cell phenotype. *Biomaterials, 31,* 1–8.

Lo, C. M., Wang, H. B., Dembo, M., & Wang, Y. L. (2000). Cell movement is guided by the rigidity of the substrate. *Biophysical Journal, 79,* 144–152.

McBeath, R., Pirone, D. M., Nelson, C. M., Bhadriraju, K., & Chen, C. S. (2004). Cell shape, cytoskeletal tension, and RhoA regulate stem cell lineage commitment. *Developmental Cell, 6,* 483–495.

Nemir, S., Hayenga, H. N., & West, J. L. (2010). PEGDA hydrogels with patterned elasticity: Novel tools for the study of cell response to substrate rigidity. *Biotechnology and Bioengineering, 105,* 636–644.

Pelham, R. J., & Wang, Y. (1997). Cell locomotion and focal adhesions are regulated by substrate flexibility. *Cell, 94,* 13661–13665.

Pouthas, F., Girard, P., Lecaudey, V., Ly, T. B. N., Gilmour, D., Boulin, C., et al. (2008). In migrating cells, the Golgi complex and the position of the centrosome depend on geometrical constraints of the substratum. *Journal of Cell Science, 121,* 2406–2414.

Rape, A. D., Guo, W.-H., & Wang, Y.-L. (2011). The regulation of traction force in relation to cell shape and focal adhesions. *Biomaterials, 32,* 2043–2051.

Singhvi, R., Kumar, A., Lopez, G. P., Stephanopoulos, G. N., Wang, D. I., Whitesides, G. M., et al. (1994). Engineering cell shape and function. *Science (New York, NY), 264,* 696–698.

Sun, Y., Jiang, L.-T., Okada, R., & Fu, J. (2012). UV-modulated substrate rigidity for multiscale study of mechanoresponsive cellular behaviors. *Langmuir: The ACS Journal of Surfaces and Colloids, 28,* 10789–10796.

Tan, J. L., Tien, J., Pirone, D. M., Gray, D. S., Bhadriraju, K., & Chen, C. S. (2002). Cells lying on a bed of microneedles : An approach to isolate mechanical force. *PNAS, 100,* 1484–1489.

Théry, M., & Piel, M. (2009). Adhesive micropatterns for cells: A microcontact printing protocol. *Cold Spring Harbor protocols, 4*(7), 888–898.

Trichet, L., Le Digabel, J., Hawkins, R. J., Vedula, S. R. K., Gupta, M., Ribrault, C., et al. (2012). Evidence of a large-scale mechanosensing mechanism for cellular adaptation to substrate stiffness. *Proceedings of the National Academy of Sciences, 109,* 6933–6938.

Tsai, I. Y., Crosby, A. J., & Russell, T. P. (2007). Surface patterning. In Y.L. Wang & D.E. Discher (Eds) *Methods in cell biology: Cell mechanics.* (pp. 67–87). San Diego, CA: Academic Press.

Tseng, Q., Wang, I., Duchemin-Pelletier, E., Azioune, A., Carpi, N., Gao, J., et al. (2011). A new micropatterning method of soft substrates reveals that different tumorigenic signals can promote or reduce cell contraction levels. *Lab on a Chip, 11,* 2231–2240.

Voskerician, G., Shive, M. S., Shawgo, R. S., von Recum, H., Anderson, J. M., Cima, M. J., et al. (2003). Biocompatibility and biofouling of MEMS drug delivery devices. *Biomaterials, 24,* 1959–1967.

Wang, H. B., Dembo, M., Hanks, S. K., & Wang, Y. (2001). Focal adhesion kinase is involved in mechanosensing during fibroblast migration. *PNAS, 98,* 11295–11300.

Wang, N., Ostuni, E., Whitesides, G. M., & Ingber, D. E. (2002). Micropatterning tractional forces in living cells. *Cell Motility and the Cytoskeleton, 52,* 97–106.

Whitesides, G. M., Ostuni, E., Takayama, S., Jiang, X., & Ingber, D. E. (2001). Soft lithography in biology and biochemistry. *Annual Review of Biomedical Engineering, 3,* 335–373.

Wong, J. Y., Velasco, A., Rajagopalan, P., & Pham, Q. (2003). Directed movement of vascular smooth muscle cells on gradient-compliant hydrogels. *Langmuir, 19,* 1908–1913.

Patterned Hydrogels for Simplified Measurement of Cell Traction Forces

2

Samuel R. Polio and Michael L. Smith

Department of Biomedical Engineering, Boston University, Boston, Massachusetts, USA

CHAPTER OUTLINE

Abstract

To understand mechanobiology, a quantitative understanding of how cells interact mechanically with their environment is needed. Cell mechanics is important to study as they play a role in cell behaviors ranging from cell signaling to epithelial to mesenchymal transition in physiological processes such as development and cancer. To study changes in cell contractile behavior, numerous quantitative measurement techniques have been developed based on the measurement of deformations of a substrate from an initial state. Herein, we present details on a technique we have developed for the measurements of 2D cellular traction forces with the goal of facilitating adaptation of this technique by other investigators. This technique is flexible in that it utilizes well-studied methods for microcontact printing and fabrication of polyacrylamide hydrogels to generate regular arrays of patterns that can be transferred onto the hydrogels. From the deformation of the arrays, an automated algorithm

ISSN 0091-679X
http://dx.doi.org/10.1016/B978-0-12-800281-0.00002-6

can be used to quantitatively determine the traction forces exerted by the cells onto the adhesion points. The simplicity and flexibility of this technique make it a useful contribution to our toolbox for measurement of cell traction forces.

INTRODUCTION

The emerging field of mechanobiology seeks to understand how cells interact mechanically with their environment. Having a quantitative understanding of how cell behavior is altered by substrate stiffness, shape, fluid shear stress, stretch or compression, and other mechanical stimuli is of great interest as we develop a deep appreciation of the mechanical nature of many diseases such as asthma and cancer (Kraning-Rush, Califano, & Reinhart-King, 2012; Kraning-Rush, Carey, Califano, & Reinhart-King, 2012; Lavoie et al., 2009; Munevar, Wang, & Dembo, 2001). It is necessary for cells to generate traction forces during processes such as individual and group cell migration, extracellular matrix (ECM) deposition, and determination of substrate stiffness (Bhadriraju et al., 2007; Discher, Janmey, & Wang, 2005; Pelham & Wang, 1997; Tambe et al., 2013). To do this, cells propagate forces via their cytoskeleton. The cytoskeleton is composed of actin and myosin fibrils, which transmit force to transmembrane proteins such as integrins and cadherins. These transmembrane ligands are able to sustain applied tension through the cell membrane as links between the filamentous actin and the ECM or other cells (Ganz et al., 2006; Stricker, Falzone, & Gardel, 2010). Depending on the types of proteins presented, the level of cell spreading and contractility may change due to altered ligand presentation, even when presented with some ligands that do not transmit force (Chopra et al., 2012; Winer, Chopra, Kresh, & Janmey, 2011). This type of behavior is known as crosstalk and is an important mechanism in understanding the mechanobiology of cells *in vivo*. To determine how various proteins contribute to the overall traction force exerted by the cell, devices must be engineered to measure such forces on multiple proteins as applied to discrete points.

Traction force systems rely on the ability to determine the deformation on a substrate in the presence of cell contractile forces, which requires image processing techniques to track these changes. One of the first ways in which cellular tractions could be semiquantitatively measured was through the observation of the contraction of cell-laden collagen hydrogels (Halliday & Tomasek, 1995). Later, the tractions exerted by individual cells were able to be measured via forces resulting in the deformation of a thin silicone sheet (Harris, Wild, & Stopak, 1980). Eventually, more quantitative techniques that permitted measurements of the deformation of soft hydrogels, for example made from polyacrylamide (PAA), were developed (Aratyn-Schaus, Oakes, Stricker, Winter, & Gardel, 2010; Kraning-Rush, Califano, et al., 2012; Kraning-Rush, Carey, et al., 2012). The degree of deformation of the substrate by the cells, which is the basis for all traction force measurement techniques, was measured by determining the displacements of beads that were randomly dispersed throughout the hydrogel (Dembo & Wang, 1999; Kraning-Rush, Califano, et al., 2012; Kraning-Rush, Carey, et al., 2012; Sniadecki & Chen, 2007). The next major advancement came with the development of micropost arrays (mPADs), which were polydimethylsiloxane (PDMS) posts whose displacement could be measured and converted to force using beam theory. Finally, micropatterning of hydrogels has been utilized for the measurement of cell traction forces through measurements of the deformation of the patterned cell contact area (Tseng et al., 2011). While each of these techniques has distinct advantages for measurement of cell traction forces, we sought to develop a new technique that is distinct both for its simplicity and high force resolution.

Herein we describe the indirect micropatterning of micron-sized, discrete fluorescent protein adhesion points onto a PAA hydrogel for the measurement of cell traction forces. This technique uses the deformation of these cell attachment points to determine traction forces based on a novel image processing algorithm that does not require user input (Polio et al., 2014). The fluorescent protein pattern is transferred to the hydrogel indirectly, enabling the use of even very soft hydrogels for this purpose. Also, techniques that have already been developed for patterning multiple proteins such as cadherins, collagen, and others onto glass substrates can be adopted to study crosstalk behavior and changes to cell contraction (Eichinger, Hsiao, & Hlady, 2012; Shen, Qi, & Kam, 2008; Shen, Thomas, Dustin, & Kam, 2008). This gives our technique an advantage as we can measure traction forces on multiple discrete, spatially distinct ligands.

2.1 METHODS

We will provide instructions on how to complete each step within the process separately in the following sections.

1. The creation of silicon masters
2. Microcontact printing
3. Coverslip activation
4. PAA gel fabrication and transfer of protein
5. Imaging and determination of cellular traction forces (CTFs).

2.1.1 The creation of silicon masters

Much of the process of the designing, creating, and troubleshooting of a silicon masters for replica molding has been covered previously in excellent, earlier editions of this book, and therefore we will only summarize the technique and highlight important differences in our approach (Sniadecki & Chen, 2007).

The first step in the experimental process is the determination of the overall design of the micropatterns that will be used to determine the CTFs. We decided to use a circular dot of 1–2 μm in diameter for our studies as this design has been used in many previous studies and techniques for determining CTFs and studying focal adhesions such as with mPADs (Balaban et al., 2001; Maloney, Walton, Bruce, & Van Vliet, 2008; Sniadecki & Chen, 2007). Also, the features had to be designed in such a manner that the aspect ratio of the feature heights to the distance between features did not exceed 5:1, nor would the posts be located close enough together (<5 μm) so that they would adhere together when removed from the mold. We have been successful with micropatterning protein dots using PDMS stamps with either 1.2 or 2 μm in diameter with a center-to-center distance of 5 or 8.45 μm, respectively, and post heights of 5 μm.

For our micropatterned dot designs, we used photolithography to create the desired features. Photolithography is the process by which UV light is used to expose a pattern on a photomask onto a surface coated with a photoreactive polymer. To create the photomask, we fabricated a chrome-on-glass master (mask acquired from Nanofilm—glass/soda lime with low reflective coating and AZ1518 photoresist). In the case of a negative photoresist, photolithography uses UV light to cure a polymer so that the unexposed area might be removed through the use of chemical agents. We used SU-8 5 (Microchem, NC0214076), which is a negative photoresist, as the photo cross-linkable polymer that produces a 5-μm thick layer of SU-8 when spun. This 5-μm tall hole then creates appropriately high posts from PDMS.

Some common issues that led to problems in pattern generation later turned out to be largely due to improper cleaning of the wafer (University Wafer, UW3P100). When spinning the SU-8, it should be smooth and not appear uneven. If the resist appears in a star-like pattern, it may be that the wafer picked up surface contaminants from the environment or was improperly cleaned. This can be rectified through Piranha cleaning the wafers. Another problem that we have experienced was that the SU-8 pattern would delaminate from the surface. This was also due to improper cleaning and the presence of surface contaminants; however, it could also be caused by underexposure of the resist to the UV light.

2.1.2 Microcontact printing

Microcontact printing is the process by which protein can be transferred from a PDMS stamp to a substrate. The process itself has been well documented and reviewed in a number of articles (Eichinger et al., 2012; Ruiz & Chen, 2007; Shen, Qi, et al., 2008; Shen, Thomas, et al., 2008; Xia & Whitesides, 1997). The technique that we will detail here is one that we have found to work well for our current laboratory setup and for our desired patterns of 1–2 μm dots in a regular array. The current procedure assumes that 25-mm coverslips are being used along with 1.5×1.5 cm^2 stamps. PDMS stamps created using photolithography and treated with 2.5% glutaraldehyde diluted in water (Sigma, G6257) and 10% 3-aminopropyltrimethoxysilane (Sigma, 281778) in the same manner as previously for the activated coverslips can be used to remove protein in unwanted areas in the desired geometry. This technique had been adapted from previous techniques to provide an element of shape control over individual or groups of cells for mechanobiology studies (Desai, Khan, Gopal, & Chen, 2011; Rottmar, Hakanson, Smith, & Maniura-Weber, 2010). It is important to note that with this technology, multiple micropatterning techniques have been developed and could be adapted for the patterning of proteins that can link chimeric proteins (e.g. protein G for linking Fc-cadherins), various ECM proteins, or other classes of proteins such as growth factors (Ganz et al., 2006; Shen, Qi, et al., 2008; Shen, Thomas, et al., 2008.

1. Mix Sylgard 184® (Dow Corning, NC9020938) in a 10:1 ratio of curing agent: base and degas under vacuum for 30 min.
2. Pour the PDMS mixture in the mold and degas under vacuum for 15 min.
3. Bake the PDMS for 2 h to cure. Do not bake for >2 h, as this can damage the mold.
4. Remove the stamp by cutting with a scalpel or sharp blade. Be cautious when removing the stamps to not crack the silicon wafer.
5. Plasma treat the PDMS stamps for 30 s using a Harrick Plasma Cleaner under air on high.
6. Add 150 µl of fluorescently labeled protein solution to the surface at a concentration of 50–100 µg/ml (Fig. 2.1A).
7. Incubate the protein solution on the stamp in a Petri dish under tin foil for 30 min to prevent photobleaching.
8. During this time, sonicate the coverslips in 95% ethanol for 10 min.
9. After incubation, tap off the protein solution and rinse lightly with 1 ml of water.
10. Allow to dry under a gentle air current such as that found in a biosafety or fume hood for 5 min.
11. After sonication, dry the coverslips under air and plasma treat under air for 1 min on high.
12. Place the stamp in contact with the coverslip as soon as the stamps have dried (Fig. 2.1B). Make sure that all of the water has evaporated, as even a small droplet causes significant patterning problems.
13. Put very light pressure on the stamp with a finger and leave it in contact with the glass for 15 min. Again, cover the stamps in foil to mitigate photobleaching.
14. During this time, it is possible to set up the second pattern with a second fluorescently labeled protein in the same manner for multiple patterns.
15. Using a piece of tape or other flat, easily visible, and low profile edge, align the edge of the first PDMS pattern while it is still on the glass with the edge of the tape (Fig. 2.1C).
16. Remove the stamp and place the second pattern in contact with the glass, making sure to align an edge with the tape (Fig. 2.1D).
17. Push slightly harder than previously and leave the stamp in contact with the glass for 30 min.

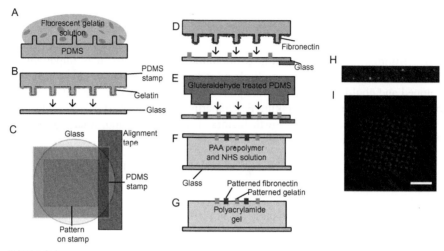

FIGURE 2.1

Microcontact printing. (A) A fluorescent gelatin solution (Molecular Probes, G13186) is added to a plasma-treated stamp and incubated for 30 min. (B) After drying, the stamp is placed on top of a plasma-treated round coverslip. (C) The edge of the pattern on the stamp is aligned with a piece of tape serving as an alignment marker before removal. (D) A second stamp is placed onto the coverslip in alignment with this marker. (E) A glutaraldehyde-treated stamp can be applied to remove protein from the unwanted areas. (F) The patterned coverslip is placed into contact with a PAA prepolymer solution. (G) Once the PAA solidifies, the coverslip can be removed and cells can be placed onto the surface. (H) A dual pattern of fibronectin (red) and gelatin (green) is patterned onto a PAA surface with two different center-to-center distances. (I) An example pattern of a patterned protein hydrogel where a glutaraldehyde-treated stamp has been used to remove the surrounding protein. The scale bar is 20 µm. (See color plate.)

18. If it is desired to confine the patterns to particular areas for studies on shape confined substrates (Fig. 2.1I), then PDMS stamps treated with glutaraldehyde and (3-aminopropyl)trimethoxysilane (3-APTMS) in the shape of the desired features can be used to remove protein by placing them in contact with the surface for 15 min (Fig. 2.1E).

19. Remove the stamp and check the pattern under a fluorescence microscope.

This process should transfer the pattern onto the coverslips. Performing micropatterning with such small features can be highly variable. In our experience, the quality of pattern transfer depends on several factors. First, the presence of excess salt can cause for crystalline patterns of protein to appear on the surface of the coverslip. This can be avoided by rinsing the stamp itself after the protein incubation, and this is also why we have decided to try to dilute protein solutions in water rather than phosphate buffered saline (PBS). Second, insufficient drying of the stamp, excessive plasma treatment, or too much pressure can cause the transfer of too much protein to the glass coverslip. It is expected that in some areas of the pattern, this will be more prevalent than others due to the way pressure is applied to the stamp. An area of sufficiently large, well-patterned protein would be an ideal pattern to use for transfer to a PAA gel. This process can require adjustments to adapt to a particular lab's equipment, so it may take some time to set the procedure up initially. This is an important control step, as poor patterns on glass will surely lead to poor patterns on the PAA gel.

The use of two patterns with different spacing between them creates a gradient pattern. This gradient is a type of Moiré pattern in which two different patterns such as grids, when overlaid on one another, create an interference pattern. In our case, we leveraged this technique to create areas of alignment using gradients as can be seen in Fig. 2.1H. This allows us to only have to align one axis of the pattern to ensure that both grids are at the same angle for appropriate alignment.

2.1.3 Coverslip activation

In this step, we make the bottom coverslip in the device chamber adhesive for PAA gel so that when the upper, patterned coverslip is removed after transfer, the hydrogel remains securely adhered to the bottom coverslip. The coverslip can be prepared at least 1 week ahead of time, which makes it convenient for preparing batches of coverslips. Similar procedures can be found elsewhere (Aratyn-Schaus et al., 2010; Kraning-Rush, Califano, et al., 2012; Kraning-Rush, Carey, et al., 2012; Rajagopalan, Marganski, Brown, & Wong, 2004).

1. Sonicate 30-mm coverslips in 98% ethanol for 10 min and then thoroughly dry using a filtered air gun. We prepare up to 6 at a time per batch in a 6-well plate (Corning, 3516).
2. Plasma treat the coverslips for 1 min using a plasma cleaner (Harrick) on high setting.
3. A very thin coat of 5% 3-APTMS in deionized (DI) water is applied to the coverslip and smeared over the surface. Excessive coating will cause an orange film to form later in the process.
4. Place the coverslips in a fume hood and allow to dry for 5 min.
5. Using DI water, thoroughly rinse the coverslips on both sides until all dried spots of 3-APTMS are removed.
6. Place coverslips into a clean 6-well plate and add approximately 2 ml of 0.5% glutaraldehyde in DI water.
7. The coverslips are incubated for 30 min and then rinsed thoroughly with water.
8. Coverslips can be dried and subsequently left in DI water for in excess of 1 week prior to use or be used immediately after drying.

2.1.4 PAA gel fabrication and transfer of protein

Once the pattern on the glass coverslip has been produced, it is important to transfer it to the hydrogel soon afterwards (<24 h) (Pelham & Wang, 1997; Rape, Guo, & Wang, 2010; Tang, Yakut Ali, & Saif, 2012). In addition to transfer to PAA gels, we have also successfully transferred patterns to flat PDMS and polymer thin films. Thus, this may be a generally accessible approach for pattern transfer onto soft substrates that are not amenable to microcontact printing.

To adjust the stiffness, the amounts of bis-acrylamide and acrylamide can be varied. The following recipe is for a total volume of pregel of 5 ml and Young's modulus of 7.6 kPa. Other formulations can be found, though it is recommended that we test the properties of the hydrogels to ensure that our formulation matches the stiffness of those reported (Kraning-Rush, Califano, et al., 2012; Kraning-Rush, Carey, et al., 2012). Note that the following procedure should be carried out in a chemical fume hood for safety, and the steps following the addition of the hydrogel should be carried out in a biosafety cabinet to help promote sterility.

1. Prepare a 40% acrylamide solution in DI water (Fisher, BP170) and add 1.25 ml to a 15 ml centrifuge tube. During this time, remove the acrylic acid N-hydroxysuccinimide (NHS)-ester from the refrigerator, as it needs to reach room temperature before opening.
2. Add 325 μl of a 2% bis-acrylamide solution in DI water (Fisher, BP171). Be careful that when introducing new solutions we do not create any bubbles or shake it. Air bubbles will cause the hydrogels to become inconsistent.
3. Add 500 μl of 10× PBS.
4. Add 2.765 ml of DI water and degas for 15 min.
5. During this time, the temperature of the acrylic acid NHS-ester (Sigma, A8060) will have been equilibrated to room temperature. Dilute the NHS-ester to 1 mg/ml in DI water and aminopropylsilane (APS; Fisher, BP179) is diluted to 100 mg/ml. APS and NHS-ester will hydrolyze over time and the activity of chemicals in the stock solution would vary, so this should be prepared new each time.
6. Add 10 μl of tetramethylethylenediamine (TEMED) (Fisher, BP150) to the prepolymer solution.
7. To decrease the pH of the solution to one which will not hydrolyze the NHS-ester, add 1 M hydrochloric acid until the pH is 7.0–7.4 (~50 μl). Gently invert the tube to ensure the solution is homogenous.
8. Add 50 μl of NHS-ester solution.
9. Add 25 μl of APS solution and invert the tube.
10. Pipette the solution immediately onto an activated coverslip within an interchangeable coverslip dish set (Bioptechs, 190310-35). The volume of solution should be 20–350 μl. This step should be carried out in a biosafety cabinet and the coverslip dish set should have been sterilized under with 70% ethanol and UV light previously.
11. Drop the coverslip on top of the solution, with the patterned side facing downward, making sure to avoid creation of air bubbles (Fig. 2.1F).
12. After 90 min, the gel will have completely polymerized. Remove the upper coverslip using a razor blade or scalpel (Fig. 2.1G). Note that it should not be left uncovered in an area with significant air flow such as a biosafety cabinet. Take care to not slide the coverslip off as this will destroy the pattern on the surface.

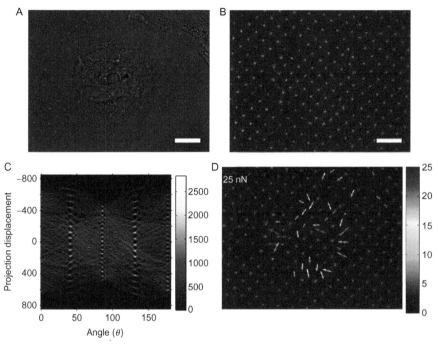

FIGURE 2.2

Analysis of cells on a patterned hydrogel. (A) A brightfield image of cells overlaid onto a fluorescent fibronectin pattern is shown. (B) The fluorescent pattern shows that a number of fibronectin adhesion points have been displaced due to the application of forces by cells. (C) The Radon transform of the image is shown. Two prominent bands are shown at approximately 40° and 130°, which represent the angle of tilt of the image. The separation between the dots represents the center-to-center distances in pixels. (D) The amount of force applied on the adhesion points is denoted by the color bar and vector length. Scale bars are 20 µm. (For interpretation of the references to color in this figure legend, the reader is referred to the online version of this chapter.)

13. To passivate remaining NHS-ester, apply 2 ml of 4% bovine serum albumin (BSA) in PBS (Fisher, BP671) to the hydrogel and incubate at 37 °C for 45 min. If desired, the gel can be briefly placed under UV light to help sterilization. The sample can be stored overnight in PBS at 4 °C until use.

After this point, cells can be added to the substrate (Fig. 2.2A). The desired cell number depends on the intended use for the substrate and whether or not the user would like to observe individual or groups of cells. Depending on cell type and the amount of protein used, the amount of time the cells will require to adhere to the surface will vary; however, it is important to recognize that sometimes, even with well-patterned substrates, cells may not appear to adhere to the desired areas. It is best to prepare several samples at once to mitigate such events and periodically observe cell adhesion to determine the behavior of our particular cells on these substrates. However, we have had success with allowing cells to adhere to the substrates overnight before washing.

2.1.5 Imaging and determination of CTFs

2.1.5.1 Imaging system

We image our cells within a customized temperature (37 °C), humidity (80%), and CO_2 (5%) regulated system to keep cells viable on the traction substrates for extended periods of time. The microscope is an Olympus IX81 with an automatic stage, a Prior Lumen200Pro light source, and Hamamatsu ORCA-R2 camera. For our filter sets, we use a set of custom Semrock filters to observe fluorescence in the Oregon Green and AlexaFluor 488 range (472/30 nm excitation and 525/30 nm emission filters) as well as the AlexaFluor 633 (628/40 nm excitation and 692/40 nm emission filters).

To accommodate thick hydrogels (>100 µm), we use a $60\times$ water lens from Olympus (numerical aperture, NA 1.00). Higher NA lenses can be used for better resolution; however, the thickness of the hydrogel can be a significant obstacle as the hydrogel must be thick enough to be modeled as having semi-infinite thickness, yet thin enough for imaging. If too thin, cells sense higher apparent substrate stiffnesses due to the underlying stiff coverslip, which would confound studies of cellular rigidity sensing (Buxboim, Rajagopal, Brown, & Discher, 2010; Maloney et al., 2008).

2.1.5.2 Image analysis and theory

To determine cellular tractions, we need to measure the deformation of the patterned substrate using the following steps:

1. Determine the location of the traction points.
2. Interpolate the initial locations of the deformed points.
3. Calculate the traction forces at each location.

We used MATLAB (Mathworks) to perform the image processing and calculations as it is a ubiquitous programming language that has many tools available to users for image processing and data analysis. The goal of the program is to determine traction forces rapidly, without any user input and preprocessing procedures that would contribute user-related errors. To access the code, we invite interested parties to visit www.bu.edu/mml/downloads. This program can also be used to calculate traction forces on mPADs by simply making small alterations to adjust for the pillar stiffness.

From the image of the deformed dot positions (Fig. 2.2B), the locations of the centroids of the traction points are determined by locating the local maxima of fluorescence. Next, a mask is created around each area with roughly the size of the maximum radius of the largest anticipated dot. From this, the centroid of the dot was found of the region, as dots can sometimes be deformed from their initial circular shape when cells apply force.

Next, the determination of the initial position of deformed points is needed. We know the initial positions of the microcontact printed points as they do not deform when properly transferred to the hydrogel (Polio, Rothenberg, Stamenović, & Smith, 2012). Since we know the initial geometric arrangement, we are able to find the dot spacing via the Radon transform of the image (Fig. 2.2C). The transform also gives the angle at which the grid is aligned at so the user does not have to be concerned with the angle of the pattern while imaging.

We use a line fitting algorithm to create perpendicular lines that are angled along each alignment axis as given from the transform. The intersections of these points create a grid, which is then used as the set of initial points from which the deformations are determined. Using a combinatorics tracking algorithm by Blair and Dufresne, we determine the origins of these points and thus their displacement (http://physics.georgetown.edu/matlab/).

Previous studies of similar systems have been performed on PDMS substrates that were micropatterned directly with fluorescent protein (Balaban et al., 2001; Stricker, Sabass, Schwarz, & Gardel, 2010). In theory, all deformations on the surface of a semi-infinite elastic substrate can be determined using the Green function, which is of the form: $u = GF$, where u is the displacements of the deformed points on the surface, G is the Green's tensor, which is defined by the material properties, and F is the force being applied by the cell at the points. To calculate the tractions, the Green's tensor must be inverted, which is not trivial due to the ill-posed nature of the problem from the noise inherent in the experimental system. Also, we wanted to minimize the amount of computation time to make it a real-time system. Therefore, we sought to use a more simplified form of the equation.

By confining the cell's focal adhesions to a circular area and assuming an infinitely thick gel, relative to the amount of deformation, we were able to apply an equation to determine traction forces as shown in Fig. 2.2D (see Maloney et al., 2008 for a full treatment of the theory). The resulting equation is very similar in nature to the techniques using the mPAD approach in that we obtain a linear relationship between the force required for the deformation of the substrate and the traction forces: $F = u\pi Ea/[(1 + v)(2 - v)]$, where F is the force vector at a point, u is the displacement vector, E is the Young's modulus of the material, and v is the Poisson's ratio of the material. In previous studies, the Poisson's ratio of PAA has ranged from 0.42 to 0.5 depending on the method of preparation and measurement technique used (Buxboim et al., 2010; Polio et al., 2012; Takigawa, Morino, Urayama, & Masuda, 1996).

One drawback of using this system is that if points are located very closely together, or points are far spaced but subjected to extreme levels of cell traction forces, the linear approximation of the forces will no longer be applicable. In the case of the points being too close (<1 μm), the traction force of neighboring points will affect one another, making it difficult to analyze the data. Extreme deformations will be plastic in nature. Therefore, it is important to take this into consideration when designing the dot spacing appropriate for the desired experiment.

2.2 EXPERIMENTAL RESULTS AND OPEN QUESTIONS

The traction force technique we present has great potential in the field of mechanobiology. Currently, we have used this technique to measure cellular tractions on PAA hydrogels with Young's moduli in the range of 2–7.6 kPa. The measured traction forces on these surfaces have ranged up to 35 nN. The lower limit of the traction forces we have determined was found experimentally by observing the traction forces measured on a well-patterned substrate that lacked any cells. Using a $60\times$ oil lens, our tracking algorithm was found to produce a roughly Gaussian distribution of tractions with the 50th percentile of dots being displaced approximately 0.094 μm (Polio et al., 2012). Experimentally, we remove displacement values below 0.3 μm to try and filter erroneous tractions that might present false positive traction values, though we may sacrifice some low force tractions. This means that our technique is limited ultimately by the pattern itself and ability to locate the centroid of the dot, as this may change as the cell deforms the substrate. We define the limit as to how accurately we can find the centroid of a pattern, as the force resolution would ultimately depend on this measurement.

The development of an open system to analyze cellular tractions on these substrates is significant and important to this technique. In our attempt to disseminate the method, we have made the code freely available and easy to use for anyone with familiarity with the MATLAB environment. The rapidness of the technique also lends it to real-time analysis while other techniques require multiple images of the experiment before and after cell removal, which necessitates postprocessing of data and would not be helpful to assess the progress of an experiment in real time.

This approach does present some drawbacks that should be mentioned. Firstly, if most of the micropatterned dots are deformed, then the analysis technique will not work well as it requires for most of the dots to be undeformed. Secondly, if the dots are located close to each other, then the deformation of one dot may interfere with another significantly. This is only an issue if the center-to-center distances of the dots are 2 μm. Thirdly, the technique is not appropriate for cell study in 3D environments; however, it would be possible to observe cells in sandwich type cultures, which would be relevant for studying cells such as hepatocytes (Beningo, Dembo, & Wang, 2004; Dunn, Yarmush, Koebe, & Tompkins, 1989). Finally, this technique limits cell adhesion to an array of dots, which is not physiologically relevant. However, other traction systems either limit cell adhesion to a uniform layer of proteins on a surface or to proteins attached to the surfaces of pillars. Cells *in vivo* attach to a fibrillar ECM, and thus all traction systems suffer from this limitation.

There has been significant interest in the ability to study how cell contractility and traction forces are altered due to the crosstalk between different ECM ligands and cell–cell adhesion molecules (Théry et al., 2005; Tseng et al., 2011). To achieve this goal, traction forces have been measured quantitatively in a number of systems, such as by micropatterning alternating bands of protein on PAA surfaces and observing cell–cell interactions on mPADs and PAA hydrogels (Borghi, Lowndes, Maruthamuthu, Gardel, & Nelson, 2010; Liu et al., 2010). However, the ability to calculate these forces depends on force balancing and is not a direct observation on the ligands involved. The ability to pattern these ligands with different geometries and the lack of a reliance on force balancing on undetermined ligands for tractions is an important feature of our technique.

With the micropatterning technique that we have developed, it is possible to align multiple patterns to achieve this effect. This allows us to determine how not only the presence and absence of particular integrins or cadherins on a surface affects cell behavior, but also how spatial resolution between ligands affects the cell's contractility (Chen, Mrksich, Huang, Whitesides, & Ingber, 1997). For example, when studying cells that exist in a 2D environment, such as endothelial cells, the ability to pattern multiple ligands would significantly enhance the study of traction forces on multiple protein surfaces (Shen, Qi, et al., 2008; Shen, Thomas, et al., 2008). There is also evidence that when interacting with multiple ECM proteins, cells will change their spreading behavior in unanticipated ways (Chopra et al., 2012; Winer et al., 2011).

It is our hope that this technology will not only be a novel part of the toolset to study mechanobiology, but will help to facilitate the interrogation of new questions that now can be asked and are of significant interest to the community. Without this technique, we could not have asked questions about the interrelated effects of ligand spacing, cell shape, and crosstalk mechanisms without this type of tool at our disposal. By providing the necessary tools and methods to conduct these experiments, we believe that this technique can become a new driving force in the study of mechanobiology.

References

Aratyn-Schaus, Y., Oakes, P. W., Stricker, J., Winter, S. P., & Gardel, M. L. (2010). Preparation of complaint matrices for quantifying cellular contraction. *Journal of Visualized Experiments, 46*, pii. 2173.

Balaban, N. Q., Schwarz, U. S., Riveline, D., Goichberg, P., Tzur, G., Sabanay, I., et al. (2001). Force and focal adhesion assembly: A close relationship studied using elastic micropatterned substrates. *Nature Cell Biology, 3*, 466–472.

Beningo, K. A., Dembo, M., & Wang, Y. L. (2004). Responses of fibroblasts to anchorage of dorsal extracellular matrix receptors. *Proceedings of the National Academy of Sciences of the United States of America, 101*, 18024–18029.

Bhadriraju, K., Yang, M., Alom Ruiz, S., Pirone, D., Tan, J., & Chen, C. S. (2007). Activation of ROCK by RhoA is regulated by cell adhesion, shape, and cytoskeletal tension. *Experimental Cell Research, 313*, 3616–3623.

Borghi, N., Lowndes, M., Maruthamuthu, V., Gardel, M. L., & Nelson, W. J. (2010). Regulation of cell motile behavior by crosstalk between cadherin- and integrin-mediated adhesions. *Proceedings of the National Academy of Sciences of the United States of America, 107*, 13324–13329.

Buxboim, A., Rajagopal, K., Brown, A. E., & Discher, D. E. (2010). How deeply cells feel: methods for thin gels. *Journal of Physics: Condensed Matter, 22*, 194116.

Chen, C. S., Mrksich, M., Huang, S., Whitesides, G. M., & Ingber, D. E. (1997). Geometric control of cell life and death. *Science, 276*, 1425–1428.

Chopra, A., Lin, V., McCollough, A., Atzet, S., Prestwich, G. D., Wechsler, A. S., et al. (2012). Reprogramming cardiomyocyte mechanosensing by crosstalk between integrins and hyaluronic acid receptors. *Journal of Biomechanics, 45*, 824–831.

Dembo, M., & Wang, Y. L. (1999). Stresses at the cell-to-substrate interface during locomotion of fibroblasts. *Biophysics Journal, 76*, 2307–2316.

Desai, R. A., Khan, M. K., Gopal, S. B., & Chen, C. S. (2011). Subcellular spatial segregation of integrin subtypes by patterned multicomponent surfaces. *Integrative Biology (Cambridge), 3*, 560–567.

Discher, D. E., Janmey, P., & Wang, Y. L. (2005). Tissue cells feel and respond to the stiffness of their substrate. *Science, 310*, 1139–1143.

Dunn, J. C., Yarmush, M. L., Koebe, H. G., & Tompkins, R. G. (1989). Hepatocyte function and extracellular matrix geometry: Long-term culture in a sandwich configuration. *FASEB Journal, 3*, 174–177.

Eichinger, C. D., Hsiao, T. W., & Hlady, V. (2012). Multiprotein microcontact printing with micrometer resolution. *Langmuir, 28*(4), 2238–2243.

Ganz, A., Lambert, M., Saez, A., Silberzan, P., Buguin, A., Mege, R. M., et al. (2006). Traction forces exerted through N-cadherin contacts. *Biology of the Cell, 98*, 721–730.

Halliday, N. L., & Tomasek, J. J. (1995). Mechanical properties of the extracellular matrix influence fibronectin fibril assembly in vitro. *Experimental Cell Research, 217*, 109–117.

Harris, A. K., Wild, P., & Stopak, D. (1980). Silicone rubber substrata: A new wrinkle in the study of cell locomotion. *Science, 208*, 177–179.

Kraning-Rush, C. M., Califano, J. P., & Reinhart-King, C. A. (2012). Cellular traction stresses increase with increasing metastatic potential. *PLoS One, 7*, e32572.

Kraning-Rush, C. M., Carey, S. P., Califano, J. P., & Reinhart-King, C. A. (2012). Quantifying traction stresses in adherent cells. *Methods in Cell Biology, 110*, 139–178.

Lavoie, T. L., Dowell, M. L., Lakser, O. J., Gerthoffer, W. T., Fredberg, J. J., Seow, C. Y., et al. (2009). Disrupting actin–myosin–actin connectivity in airway smooth muscle as a treatment for asthma? *Proceedings of the American Thoracic Society, 6*, 295–300.

Liu, Z. J., Tan, J. L., Cohen, D. M., Yang, M. T., Sniadecki, N. J., Ruiz, S. A., et al. (2010). Mechanical tugging force regulates the size of cell-cell junctions. *Proceedings of the National Academy of Sciences of the United States of America, 107*, 9944–9949.

Maloney, J. M., Walton, E. B., Bruce, C. M., & Van Vliet, K. J. (2008). Influence of finite thickness and stiffness on cellular adhesion-induced deformation of compliant substrata. *Physics Review E, 78*, 041923.

Munevar, S., Wang, Y., & Dembo, M. (2001). Traction force microscopy of migrating normal and H-ras transformed 3T3 fibroblasts. *Biophysics Journal, 80*, 1744–1757.

Pelham, R. J., Jr., & Wang, Y. (1997). Cell locomotion and focal adhesions are regulated by substrate flexibility. *Proceedings of the National Academy of Sciences of the United States of America, 94*, 13661–13665.

Polio, S. R., Parameswaran, H., Canović, E. P., Gaut, C. M., Aksyonova, D., Stamenović, D., et al. (2014). Topographical control of multiple cell adhesion molecules for traction force microscopy. *Integrative Biology*, http://dx.doi.org/10.1039/C3IB40127H.

Polio, S. R., Rothenberg, K. E., Stamenović, D., & Smith, M. L. (2012). A micropatterning and image processing approach to simplify measurement of cellular traction forces. *Acta Biomaterialia, 8*, 82–88.

Rajagopalan, P., Marganski, W. A., Brown, X. Q., & Wong, J. Y. (2004). Direct comparison of the spread area, contractility, and migration of balb/c 3T3 fibroblasts adhered to fibronectin- and RGD-modified substrata. *Biophysics Journal, 87*, 2818–2827.

Rape, A. D., Guo, W. H., & Wang, Y. L. (2010). The regulation of traction force in relation to cell shape and focal adhesions. *Biomaterials, 32*, 2043–2051.

Rottmar, M., Hakanson, M., Smith, M., & Maniura-Weber, K. (2010). Stem cell plasticity, osteogenic differentiation and the third dimension. *Journal of Materials Science: Materials in Medicine, 21*, 999–1004.

Ruiz, S. A., & Chen, C. S. (2007). Microcontact printing: A tool to pattern. *Soft Matter, 3*, 168–177.

Shen, K., Qi, J., & Kam, L. C. (2008). Microcontact printing of proteins for cell biology. *Journal of Visualized Experiments, 22*, pii: 1065.

Shen, K., Thomas, V. K., Dustin, M. L., & Kam, L. C. (2008). Micropatterning of costimulatory ligands enhances CD4+ T cell function. *Proceedings of the National Academy of Sciences of the United States of America, 105*, 7791–7796.

Sniadecki, N. J., & Chen, C. S. (2007). Microfabricated silicone elastomeric post arrays for measuring traction forces of adherent cells. *Methods in Cell Biology, 83*, 313–328.

Stricker, J., Falzone, T., & Gardel, M. L. (2010). Mechanics of the F-actin cytoskeleton. *Journal of Biomechanics, 43*, 9–14.

Stricker, J., Sabass, B., Schwarz, U. S., & Gardel, M. L. (2010). Optimization of traction force microscopy for micron-sized focal adhesions. *Journal of Physics: Condensed Matter, 22*, 194104.

Takigawa, T., Morino, Y., Urayama, K., & Masuda, T. (1996). Poisson's ratio of polyacrylamide (PAAm) gels. *Polymer Gels and Networks, 4*, 1–5.

Tambe, D. T., Croutelle, U., Trepat, X., Park, C. Y., Kim, J. H., Millet, E., et al. (2013). Monolayer stress microscopy: Limitations, artifacts, and accuracy of recovered intercellular stresses. *PLoS One, 8*, e55172.

Tang, X., Yakut Ali, M., & Saif, M. T. A. (2012). A novel technique for micro-patterning proteins and cells on polyacrylamide gels. *Soft Matter, 8*, 3197–3206.

Théry, M., Racine, V., Pepin, A., Piel, M., Chen, Y., Sibarita, J. B., et al. (2005). The extracellular matrix guides the orientation of the cell division axis. *Nature Cell Biology, 7*, 947–953.

Tseng, Q., Wang, I., Duchemin-Pelletier, E., Azioune, A., Carpi, N., Gao, J., et al. (2011). A new micropatterning method of soft substrates reveals that different tumorigenic signals can promote or reduce cell contraction levels. *Lab on a Chip, 11*, 2231–2240.

Winer, J. P., Chopra, A., Kresh, J. Y., & Janmey, P. A. (2011). Chapter 2: Substrate elasticity as a probe to measure mechanosensing at cell-cell and cell-matrix junctions. In A.W. Johnson & B. Harley (Eds.), *Mechanobiology of Cell-Cell and Cell-Matrix Interactions* (pp. 11–22).

Xia, Y. N., & Whitesides, G. M. (1997). Extending microcontact printing as a microlithographic technique. *Langmuir, 13*, 2059–2067.

Micropatterning Hydroxy-PAAm Hydrogels and Sylgard 184 Silicone Elastomers with Tunable Elastic Moduli

3

Marie Versaevel[1], Thomas Grevesse[1], Maryam Riaz, Joséphine Lantoine, and Sylvain Gabriele

Mechanobiology and Soft Matter Group, Laboratoire Interfaces et Fluides Complexes, Complexys and Biosciences Research Institutes, CIRMAP, Université de Mons, Mons, Belgium

[1]*These authors contributed equally*

CHAPTER OUTLINE

ISSN 0091-679X
http://dx.doi.org/10.1016/B978-0-12-800281-0.00003-8

Abstract

This protocol describes a simple method to deposit protein micropatterns over a wide range of culture substrate stiffness (three orders of magnitude) by using two complementary polymeric substrates. In the first part, we introduce a novel polyacrylamide hydrogel, called hydroxy-polyacrylamide (PAAm), that permits to surmount the intrinsically nonadhesive properties of polyacrylamide with minimal requirements in cost or expertize. We present a protocol for tuning easily the rigidity of "soft" hydroxy-PAAm hydrogels between ~0.5 and 50 kPa and a micropatterning method to locally deposit protein micropatterns on these hydrogels. In a second part, we describe a protocol for tuning the rigidity of "stiff" silicone elastomers between ~100 and 1000 kPa and printing efficiently proteins from the extracellular matrix. Finally, we investigate the effect of the matrix rigidity on the nucleus of primary endothelial cells by tuning the rigidity of both polymeric substrates. We envision that the complementarity of these two polymeric substrates, combined with an efficient microprinting technique, can be further developed in the future as a powerful mechanobiology platform to investigate *in vitro* the effect of mechanotransduction cues on cellular functions, gene expression, and stem cell differentiation.

INTRODUCTION

It is now widely accepted that the mechanical environment plays an important role in directing cell fate and tissue morphogenesis (Guilak et al., 2009; Lecuit & Lenne, 2007). Several reports have demonstrated that cells undergo drastic morphological changes in response to external forces (e.g., shear stress, compression, or stretch) and modifications of the physicochemical properties of the extracellular matrix (ECM) (e.g., durotaxis, chemotaxis, or haptotaxis). The cell microenvironment presents narrow pores that impose a confined adhesiveness, which in turn influences the cellular shape and cytoskeletal organization (Gabriele, Benoliel, Bongrand, & Théodoly, 2009; Gabriele, Versaevel, Preira, & Théodoly, 2010; Pathak & Kumar, 2012). These stimuli are translated in adherent cells into a cascade of biochemical events through complex mechanotransduction pathways (Jaalouk & Lammerding, 2009). In this context, recent works have suggested the existence of a mechanical coupling between the ECM and the nucleus (Versaevel, Grevesse, & Gabriele, 2012;

Versaevel, Grevesse, Riaz, & Gabriele, 2013; Wang, Tytell, & Ingber, 2009) that may explain how modifications of the ECM can alter the gene machinery. The elucidation of the signaling pathways underlying the regulation of cellular shapes and nuclear functions requires the identification of intracellular components that are specifically activated in response to different ECM cues.

To tackle this problem, various natural and synthetic *in vitro* platforms have been refined and created to reproduce the complexity of the ECM. Indeed, most of the experiments in the field of mechanobiology were performed until recently in plastic dishes, which represent a physiologically inappropriate environment. As the importance of the mechanical properties of the ECM has been realized, researchers have started to develop synthetic materials that can recapitulate the range of physiological ECM rigidities. Some of these works have revealed that migration of 3T3 fibroblasts and rat kidney epithelial cells is regulated by the ECM stiffness (Pelham & Wang, 1997), that fibroblasts tune their internal stiffness to match that of the ECM (Solon, Levental, Sengupta, Georges, & Janmey, 2007), and that changes in the ECM stiffness can drive the differentiation of mesenchymal stem cells into specific lineages (Engler, Sen, Sweeney, & Discher, 2006).

By combining these novel materials with elegant microprinting technologies, biologically relevant questions have been recently assessed via the creative design of well-defined microenvironments. For instance, researchers have sought to develop a diverse set of complex microenvironments to probe the complexity of cell–substrate interactions in cellular motility (Brock et al., 2003; Théry, 2010), differentiation (Gao, McBeath, & Chen, 2010), or cell division (Kwon et al., 2008; Théry et al., 2005). Among these works, many efforts have been directed toward understanding the relationship between cell shape and nuclear functions (Thomas, Collier, Sfeir, & Healy, 2002; Versaevel et al., 2012, 2013) and decoupling the effect of the matrix stiffness on nuclear mechanotransduction (Grevesse, Versaevel, Circelli, Desprez, & Gabriele, 2013).

Most of cell types exhibit an elastic modulus in the range of $\sim$1 to $\sim$100 kPa, but there are also some tissues that experience a stiffer environment (Palchesko, Zhang, Sun, & Feinberg, 2012), such as cardiac muscle tissues ($\sim$40–400 kPa), arterial walls ($\sim$800 kPa), or basement membranes ($\sim$1–3 MPa). However, only very few materials have been reported to be able to cover the entire range of soft elastic moduli (from kPa to MPa) that can be found in the human body. To address this challenge, we describe how microcontact printing can be easily combined with modified polyacrylamide hydrogels for creating modified "soft" (0.5–50 kPa) culture substrates and silicone elastomers for modified "stiff" ($\sim$100 kPa–1 MPa) matrices (Fig. 3.1).

3.1 PHOTOMASK AND MICROSTAMPS

The microcontact printing technique requires the design of a photomask in a drawing software that can be open source (e.g., FreeCAD or LibreCAD) or commercial (e.g., Clewin or Autocad). These softwares allow to draw mask features with a micrometer

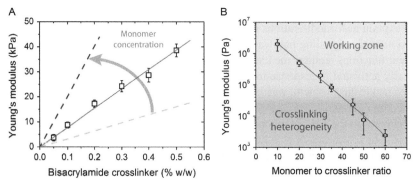

FIGURE 3.1

Elasticity of hydroxy-PAAm hydrogels and silicone gels. (A) Hydroxy-PAAm hydrogel elasticity versus bis-acrylamide cross-linker concentration (in w/w %). The linear fit (red line) shows the linear evolution of the elasticity with the cross-linker ratio, whereas dashed blue lines (light and dark) indicate that the range of elastic moduli can be adjusted by changing the monomer concentration. (B) The linear evolution of the Young's modulus of the Sylgard 184 as a function of monomer to cross-linker ratio indicates that varying the cross-linking density of Sylgard 184 allows to tailor the mechanical properties of the substrate. Variability in the Young's modulus increased in very soft samples (from 40:1 to 60:1) and may have resulted from cross-linking heterogeneity. The green region corresponds to the working area, whereas the red region corresponds to the presence of cross-linking heterogeneities in the material, which are associated with a sticky behavior. (For interpretation of the references to color in this figure legend, the reader is referred to the online version of this chapter.)

size resolution and then to export planar design geometries in a specific file format for the mask manufacturer, such as the Graphic Database System format (common acronym, GDS II).

Based on this design, a chrome-coated photomask in transparent fused silica or synthetic quartz is fabricated in a clean room. This photomask will permit to expose to ultraviolet (UV) light-specific regions of a thin layer of photoresist, beforehand spin coated onto a flat silicon substrate. The surface of the silicon substrate must be cleaned carefully to avoid the presence of defects that may induce the dewetting of the photoresist layer (Hamieh et al., 2007; Reiter et al., 2009). After UV photo-illumination, the soluble photoresist is washed off using a liquid developer in order to obtain the photomask pattern made of cross-linked photoresist. At this stage, the silicon wafer coated with a thin layer of photoresist can be used as a topographic template for molding polydimethylsiloxane (PDMS) stamps (Fig. 3.2A). Alternatively, unprotected regions of the silicon master can be anisotropically etched, for instance by deep reactive-ion etching and then the photoresist stripped in order to obtain a more stable silicon template. The silicon master must be finally treated with the vapor of octadecyltrichlorosilane (or a perfluorosilane such as $C_6FI_3C_2H_5SiCl_3$) to cap

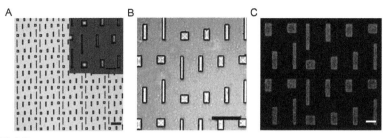

FIGURE 3.2

From the silicon master to protein micropatterns. (A) Optical micrograph of the surface of a silicon master structured with square-shaped and rectangular microscopic features. The inset shows a small zone of the silicon master. Scale bars are 200 μm for the master and 100 μm for the inset. (B) Optical micrograph of the surface of a silicone stamp obtained from the molding of the silicon master presented in (A). The scale bar is 100 μm. (C) Optical fluorescence image of different geometries of fibronectin micropatterns deposited by microcontact printing on a PDMS-coated coverslip ($E = 1$ MPa). The scale bar is 50 μm. (For color version of this figure, the reader is referred to the online version of this chapter.)

any polar reactive groups (e.g., –OH) present at the surface of the master with a chemically inert group such as –CH$_3$ or –CF$_3$.

A sylgard 184 prepolymer, which is a commercial silicon elastomer, with a 1:10 ratio of curing agent to base, is poured onto the silanized silicon master and then cured at 60 °C during 3 h to cross-link the polymer. Finally, the PDMS layer is peeled off from the template and stamps with a dimension of 10×10 mm^2 are cut with a scalpel. The stamps used for μCP are made of PDMS to mold with very high fidelity the microfeatures of the silicon master (Fig. 3.2B). Due to its elastomeric behavior, a PDMS stamp will deform enough to conform to the surface (Fig. 3.2C). In addition, the low surface energy of PDMS ($\sim$23 mJ/m^2) facilitates its separation from the silicon master and also to peel it easily from the printed surface.

3.2 MICROSTAMP FEATURES

Micropatterns were designed for studying at the single-cell level the effect of cell shape changes on the behavior of the nucleus. *In vivo*, endothelial cells reorientate and elongate in the direction of the blood shear stress. Low and oscillatory shear stress, that can be found in bifurcated and curved vessels, have been shown to be associated with a cobblestone-like shape of endothelial cells, whereas high and unidirectional shear stress were observed to induce elongated endothelial morphologies, up to an aspect ratio of 1:10. Consistent with these observations, we designed square (aspect ratio 1:1) and rectangular micropatterns with a wide range of aspect ratios (from 1:2 to 1:10) in order to reproduce *in vitro*, at the single-cell level, the different morphologies adopted *in vivo* by endothelial cells (Figs. 3.2C).

Interestingly, previous reports have demonstrated that the spreading area of adherent cells is a key parameter of the cell–substrate interactions (Yeung et al., 2005), suggesting that the spreading area is a parameter involved in the generation of intracellular forces that control the nuclear shape. Consistent with this observation, it can be therefore useful to investigate the role of the cellular morphology on the nuclear shape remodeling process by controlling the cell surface area. Previous reports have shown that mammalian cells need enough space to spread, whereas low spreading areas lead to apoptosis (Chen, Mrksich, Huang, Whitesides, & Ingber, 1997). In this work, each micropattern geometry was designed to adopt five values of areas: 600, 800, 1200, 1600, and 2000 μm^2 in order to decipher the role of the cell spreading area in the nuclear remodeling process. It is interesting to note that a gap of 50 μm was placed between micropatterns to prevent cell spreading on neighboring patterns.

3.3 SOFT MATRICES: HYDROXY-PAAm HYDROGELS

3.3.1 Materials

3.3.1.1 Reagents

Acrylamide in powder (Sigma-Aldrich, A3553)
N,N'-methylenebis(acrylamide) (Sigma-Aldrich, 146072)
N-hydroxyethylacrylamide (Sigma-Aldrich, 697931)
N,N,N',N'-tetramethylenediamine (Sigma-Aldrich, T9281)
Sodium hydroxide (Sigma-Aldrich, 221465-25G)
Ammonium persulfate (APS) (Sigma-Aldrich, A3678)
Double-distilled water, ddH20
3-(trimethoxysilyl)propyl acrylate (Sigma-Aldrich, 1805)
Human plasma fibronectin (Millipore, FC010)
Laminin (Sigma-Aldrich, L2020)
Sterile phosphate buffered saline (PBS) (PAA Laboratories, H15-002).

3.3.1.2 Equipments

Laminar flow hood (Filtest Clean Air Technology)
Ultrasonic bath tray (Sigma-Aldrich, Z613983-1EA)
Vortexer (Scientific Industries, Vortex Genie 2)
Rocking plate (IKAcWerke, KS 130 Basic)
UV/Ozone Photoreactor (Ultraviolet Company, Model PR-100)
Vacuum degassing chamber (Applied Vacuum Engineering, DP-8-Kit)
Parafilm (Sigma-Aldrich, P7793-1EA)
Dressing tissue forceps (Sigma-Aldrich, F4391-1EA)
Polystyrene petri dishes (Sigma-Aldrich, P5731-500EA)
0.2 μm Isopore membrane filter (Millipore, GTTP Filter code)
22 mm round glass coverslip (Neuvitro, GG-22)
25 mm round glass coverslip (Neuvitro, GG-25)
Variable volume micropipette (Sigma-Aldrich, Z114820)

Microcentrifuge tubes (Sigma-Aldrich, Z666505-100EA)
75 cm^2 cell culture flask (Sigma-Aldrich, CLS430641)
Aluminum foil (Sigma-Aldrich, 266574-3.4G).

3.3.2 Method

3.3.2.1 Step #1: hydroxy-PAAm hydrogel fabrication

1. Place 25-mm-diameter glass circular coverslips in a petri dish and under a chemical fume hood smear 0.1 M NaOH solution on it for 5 min.
2. Drain the NaOH solution and fully immerse coverslips in sterile ddH$_2$O for 20 min under a gentle rocking.
3. Drain sterile ddH$_2$O and repeat step 2.
4. Remove coverslips with fine tweezers and place them in a new petri dish with the activated face up by squeezing a drop of sterile ddH$_2$O between the glass coverslip and the plastic petri dish.
5. Dry coverlips under a steady flow of high-purity nitrogen gas. The drop of sterile ddH$_2$O squeezed between the petri dish and the coverslip will avoid the coverslip to move.
6. In a culture hood, smear a thin layer of 3-(trimethoxysilyl)propyl acrylate (92%) on the activated side of the dried coverslips for 1 h.
7. Wash glass coverslips with three washes of sterile ddH$_2$O and then immerse them with sterile ddH$_2$O in a new petri dish.
8. Tap the petri dish with parafilm and place it on a rocker plate under gentle agitation for 10 min.
9. Remove the coverslips from ddH$_2$O with sterile tweezers and place them in a new petri dish with the activated face up.
10. Prepare 65 mg of *N*-hydroxyethyl acrylamide (HEA) in a 1.5 mL Eppendorf tube.
11. Add 1 mL of 50 mM sterile HEPES buffer to HEA and mix using a vortexer until the complete HEA dissolution.
12. Add 400 µL of 40% w/w in HEPES acrylamide solution and the required volume of 2% w/w in HEPES bis-acrylamide solution to reach the desired hydrogel stiffness (see Fig. 3.1A). Adjust with 50 mM HEPES to reach a final volume of 5 mL.
13. Mix the solution using a vortexer during 2 min and degas it in a vacuum chamber for 20 min. This step will permit to reduce the oxygen concentration within the solution that would prevent the polymerization of the hydroxy-PAAm solution.
14. Under a sterile hood, sterilize the degassed solution with a 0.2 µm pore size filter.
15. Activate 22-mm-diameter circular glass coverslips in a UV/Ozone cleaner during 7 min.
16. Prepare 100 µL of 10% APS solution, that is, 10 mg APS in 100 µL ddH$_2$O.

17. Add 2.5 μL of tetramethylenediamine (TEMED) and 25 μL of APS solution to the sterilized hydroxy-PAAm solution to initiate the polymerization. Under sterile conditions, mix the solution by three successive pipettings without introduction of bubbles.
18. Under a sterile hood, place a 25 μL drop of the hydroxy-PAAm solution on a 25-mm coverslip (from step 9) and immediately place a 22-mm glass coverslip (from step 15) on top of the droplet to squeeze the hydroxy-PAAm solution. Center the 22-mm glass coverslip with sterile tweezers and smooth out any bubbles.
19. Let hydroxy-PAAm hydrogels polymerize at room temperature for 15–30 min. Invert manually the remaining hydroxy-PAAm solution in the Eppendorf tube to follow the completion of the polymerization process.
20. Immerse coverslips in sterile ddH$_2$O and carefully separate the 22-mm glass coverslips by introducing the edge of a razor blade between the 22-mm glass coverslips and the hydroxy-PAAm hydrogel layer.
21. Wash with hydroxy-PAAm coated coverslips with three washes of sterile PBS.
22. Store hydroxy-PAAm hydrogels in sterile PBS at 4 °C for up to 3 days.

3.3.2.2 Step #2: Microcontact printing on hydroxy-PAAm hydrogels

1. Wash the PDMS microstamps by sonicating them for 15 min in a 70% solution of ethanol in water.
2. Dry the stamps with filtered nitrogen and place them in a petri dish with their microfeatures head-up.
3. Place the opened petri dish in a UV/ozone cleaner ($\lambda < 200$ nm) for 7 min.
4. Close the petri dish and bring it into the laminar flow hood.
5. Under sterile conditions, prepare a 25 μg/mL solution of fibronectin (or for instance a 100 μg/mL of laminin in PBS) in sterile water and spread 200 μL of this solution on the top of each stamp by moving it with a sterile tip of a pipette toward each corner of the stamp. Do not mix the fibronectin solution thoroughly.
6. Let the protein solution incubate on the stamps for 1 h at room temperature under the sterile hood. Turn off lamps to avoid protein damage.
7. Under a sterile hood, transfer hydroxy-PAAm coated coverslips into a new petri dish and remove excess PBS from the surface with a low nitrogen stream. Stop the procedure as soon as no evidence of standing water on the gel surface is observed. The gel should not be dried thoroughly at this stage.
8. Place the structured surface of the stamp in contact with the dried hydrogel surface. Ensure a good contact between the stamp and the hydrogel surface.
9. Leave the PDMS stamp on the hydrogel surface for 1 h at room temperature.
10. Gently remove PDMS stamps from hydroxy-PAAm hydrogels with forceps.
11. Under sterile conditions, wash extensively stamped hydroxy-PAAm hydrogels by three exchanges of PBS (pH 7.4) for 10 min per exchange and place them in a 6-well petri dish.

12. Passivate nonprinted zones by adding 3 mL/well of a sterile solution of Bovine Serum Albumin (BSA) at 5 mg/mL in PBS during one night at 4 °C under a gentle agitation on a rocking plate.
13. Wash extensively by three exchanges of PBS (pH 7.4) in sterile conditions for 10 min per exchange. At this stage, stamped hydroxy-PAAm hydrogels can be stored at 4 °C up to 1 week.

3.4 STIFF MATRICES: SILICONE ELASTOMERS

In this section, we describe the preparation of silicone elastomers with tunable rigidity and the procedure for micropatterning PDMS-coated coverslip with proteins from the ECM.

3.4.1 Materials
3.4.1.1 Reagents
   Sylgard 184 silicone elastomer kit (Dow Corning)
   Pluronic F127 Micropastille (BASF)
   Petri dishes (Greiner bio one, 633102)
   6-well culture plates (Greinerbio one, 657160)
   25-mm-diameter glass coverslips (VWR, 631-1584)
   Ethanol 99.8% (Sigma Aldrich, 02860)
   Fibronectin from human plasma (Millipore, FC010)
   Sterile PBS (PAA Laboratories, H15-002).

3.4.1.2 Equipments
   UV/Ozone Photoreactor (Ultraviolet Products, model PR-100)
   Rocking plate (IKAcWerke, Model KS 130 Basic)
   Conditioning mixer (Thinky, ARE-250, USA)
   Wafer spinner (Polos, MDC 300, The Netherlands)
   Laminar flow hood (Filtest Clean Air Technology)
   Ultrasonic bath (Sonoclean).

3.4.2 Method
3.4.2.1 Step #1: Fabrication of PDMS-coated coverslips
1. Round glass coverslips are cleaned by a 15-min sonication in a 70% solution of ethanol in water and dried with filtered nitrogen.
2. Prepare Sylgard 184 elastomer by mixing the PDMS monomer and its curing agent. A final elastic modulus ranging from 5×10^4 and 3×10^6 Pa can be obtained by adjusting the ratio (monomer:curing agent) between 40:1 and 10:1 (see Fig. 3.1B). In order to ensure a good mixing and degassing, the uncured

     solution is placed in a conditioning mixer for 10 min. Alternatively, the mixture can be mixed manually during 10 min and then degassed under vacuum for 2 h.

3. Place a glass coverslip on a spin coater, then cover it with a drop of uncured PDMS elastomer and start the spin-coating program. In order to optimize the quality of the PDMS layer deposited, spin-coat with a speed gradually increasing from 500 to 5000 rpm, for a total duration of 2 min. Deposit the spin-coated coverslips in a petri dish.

4. Cure the PDMS layer by placing the petri dish containing the coated coverslips 3 h at 60 °C in an oven, regardless the value of the elastic modulus. Alternatively, let the coverslips cure for 24 h at room temperature.

3.4.2.2 Step #2: Microcontact printing on PDMS-coated coverslips

1. Wash the PDMS microstamps by sonicating them for 15 min in a 70% solution of ethanol in water.

2. Dry the stamps with filtered nitrogen and place them in a petri dish with their microfeatures head-up.

3. Place the opened petri dish in a UV/ozone cleaner ($\lambda < 200$ nm) oven for 7 min.

4. Close the petri dish and bring it into the laminar flow hood.

5. Under sterile conditions, prepare a 25 µg/mL solution of fibronectin (or any other ECM proteins such as laminin or collagen) in sterile water and spread 400 µL of this solution on the top of each stamp. Do not mix the fibronectin solution thoroughly.

6. Let the fibronectin solution incubate on the stamps for 1 h at room temperature under the sterile hood.

7. Prepare a 1% w/w solution of Pluronic F 127 in water. Filter this solution with a 0.20-µm pore size filter in order to sterilize it.

8. About 10 min before the end of the incubation time, place the PDMS-coated coverslips in the UV/ozone and treat them for 7 min. Close the petri dish and bring it into the laminar flow hood.

9. Dry the stamps with filtered nitrogen under the laminar flow hood.

10. Place a stamp head-down on an activated PDMS-coated coverslip for 30 s.

11. With sterile forceps, remove the stamp from the coverslip and place the stamped coverslips in a 6-well culture plate.

12. Add 4 mL of the Pluronic solution on the wells during 5 min in order to passivate the unfunctionalized regions of the coverslip.

13. Rinse three times with 4 mL of sterile PBS. The microcontact printed coverslips can be used immediately or stored in sterile PBS at 4 °C for up to 1 week.

3.5 CELL DEPOSITION

This part describes the method used to control the shape of single Human Umbilical Vein Endothelial Cells (HUVECs) by deposing protein micropatterns on PDMS surfaces, as shown in Versaevel et al. (2012), and hydroxy-PAAm surface, as shown in

Grevesse et al. (2013). These methods have been used successfully in our laboratory with the following cell types: primary cortical neurons, normal human dermal fibroblasts, ECV 304, bone osteosarcoma cells (MG-63 cell line), human osteosarcoma-derived cells (SaOs-2), human osteoprogenitor (HOP), and human skeletal myoblasts (LHCN M2).

3.5.1 Materials

3.5.1.1 Reagents
PBS (PAA Laboratories, HIS-002)
Trypsin (0.5 g/L)–ethylenediaminetetraacetic acid (EDTA) (0.2 g/L) (Invitrogen/Gibco, 25300-054)
Endothelial cell growth medium (Cell Applications, 211 K-500)
Fetal bovine serum (PAA Laboratories, AIS_ISI)
Penicillin–streptomycin (Invitrogen/Gibco, 15140-122)
HUVECs (Invitrogen, C-003-5C)
Absolute ethanol (Sigma-Aldrich, 459844-2.5 L).

3.5.1.2 Equipments
Laminar Flow hood (Filtest Clean Air Technology)
75-cm^2 cell culture flask (Sigma-Aldrich, CLS430641)
Protein microcentrifuge tubes (Sigma-Aldrich, Z666505-100EA)
Cell incubator (Binder, APT line C150)
Centrifuge (Hermle, Z 252 MK)
Stainless steel forceps with fine tips (Sigma Aldrich, Z225304-1EA)
Cell culture multidish six wells (Sigma Aldrich, Z688649-4CS)
Variable volume micropipette (Sigma-Aldrich Z114820)
Sterile pipettes (Sigma-Aldrich, CLS70785N).

3.5.2 Method

1. Incubate PDMS or hydroxy-PAAm-coated coverslips in sterile culture medium at 37 °C for 30–45 min prior to plating cells.
2. Wash adherent cells cultured in a 75-cm^2 culture flask with sterile PBS at 37 °C and then detach cells with 3 mL of trypsin–EDTA or accutase for 10 min.
3. Transfer the desired amount of prewarmed complete growth medium appropriate for your cell line into the flask containing the detached cells and transfer the cell suspension into a tube for centrifugation.
4. Centrifuge the cell suspension for 3 min at 2000 rpm.
5. Under a sterile hood, remove the supernatant with a micropipette and resuspend the cells in complete culture medium (37 °C) at 15–20,000 cells/mL.
6. Place coverslips obtained from Step #1 in a cell culture multidish.
7. Add 4 mL of the cell solution in a well and place the multidish plate in a culture incubator at 37 °C and 5% humidity for 1–2 h.

8. Under a sterile hood, gently aspirate unattached cells and replace the culture medium with a fresh warmed one.
9. Return the attached cells to the incubator and let them spread onto the micropatterned surface (3–6 h, depending on the cell type).

3.6 DISCUSSION

Several types of polymer can be used to fabricate matrix substrates of varying stiffness. Here, we introduce two simple methods to create matrices of tunable elasticity by using hydroxy-PAAm hydrogels and Sylgard 184 silicone elastomers. Both polymers offer several important advantages for cell culture, such as constant surface chemistry in their working range of mechanical properties, translucent quality for optical and fluorescence microscopy, and compatibility with the micropatterning technique. As shown in Fig. 3.3, the complementarity of "soft" hydroxy-PAAm hydrogels and "stiff" Sylgard 184 silicone elastomers allows to obtain a three order-of-magnitude range of tunability that permits to recapitulate the whole range of stiffnesses (from ~1 kPa to ~1 MPa) observed for cells and tissues.

By combining the microcontact printing technique with both materials, we have investigated the role of the matrix stiffness on the regulation of the nucleus. Indeed, orientation and deformation of the nucleus are regulated by lateral compressive forces, which are driven by accumulated tension in central actin fibers (Versaevel et al., 2012). By changing the substrate stiffness over three orders of magnitude,

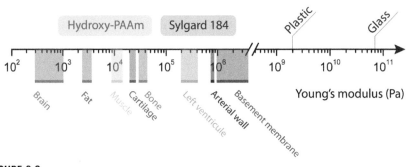

FIGURE 3.3

Soft tissue elasticity scale ranging from soft brain, fat, striated muscle, cartilage, and precalcified bone to left ventricule at peak systole and arterial walls. In contrast, conventional culture surfaces, such as plastic dishes and glass coverslips, exhibit much stiffer elastic moduli (~2 × 10^9 and ~7 × 10^{10} Pa, respectively). The top part of the panel shows the range of elastic moduli available with hydroxy-PAAm hydrogels (in green) and the sylgard 184 silicone elastomer (in blue). (For interpretation of the references to color in this figure legend, the reader is referred to the online version of this chapter.)

we have studied whether significant modifications of the matrix stiffness can impact the nuclear shape by changing the level of tension in actin stress fibers. Considering that large variations of the spreading area have been observed previously in response to matrix stiffness modifications, we plated single HUVECs on rectangular fibronectin (FN)-coated micropatterns with a constant area of 1200 μm². This procedure permits to control in the same assay spreading and morphology of single HUVECs. These FN micropatterns were deposited on hydroxy-PAAm hydrogels of 2.5, 8.5, and 25 kPa and silicone elastomers of 1 MPa. It is important to note that fluorescent assays performed on these four substrates have shown a similar surface chemistry, regardless of the polymer nature and the rigidity value. Therefore, by controlling (i) the cellular shape, (ii) the spreading area, and (iii) the protein coating, we ensure that any eventual modifications of the nuclear shape will be directly related to changes in the matrix stiffness. After 24 h in culture, HUVECs were fixed and immunostained to observe the spatial distribution of actin filaments, focal adhesions, and the shape of the nucleus (Fig. 3.4).

We observed that cells plated on stiffer substrates had thicker and straighter actin fibers, a more deformed nucleus, and well-defined focal adhesion areas. These observations indicate that modifications of the cell matrix stiffness can impact the nucleus without changing the cellular shape, the cell–substrate adhesion, or the spreading area.

This suggests that changes of matrix rigidity modulate the production of internal tension in actin fibers, which result in a remodeling of the nucleus. By using morphometric analysis of the nucleus, we demonstrated that the projected nuclear area decreased significantly with the substrate stiffening. As shown in Fig. 3.5A, we observed a decrease of ~52% from 2.5 kPa to 1 MPa. Then, we quantified the nuclear

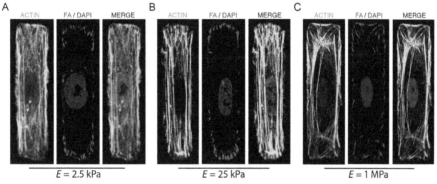

FIGURE 3.4

Immunostained HUVECs cultured on rectangular FN-coated micropatterns (aspect ratio 1:4) deposited on (A) $E = 2.5$ kPa, (B) $E = 25$ kPa hydroxy-PAAm hydrogels, and (C) 1 MPa silicone elastomers. (See color plate.)

Reproduced from Grevesse et al. (2013) by permission of The Royal Society of Chemistry.

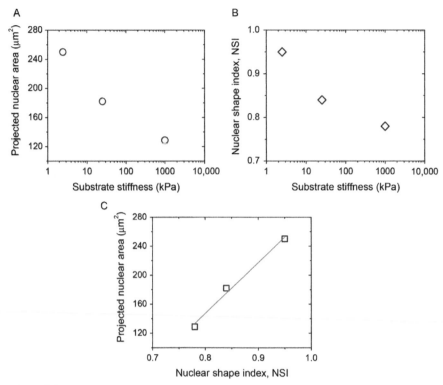

FIGURE 3.5

Morphometric evolution of the nucleus of single HUVECs cultured on rectangular-FN micropatterns (aspect ratio 1:4) that were deposited on 2.5 and 25 kPa hydroxy-PAAm hydrogels and 1 MPa silicone elastomers. Evolution of (A) the projected nuclear area versus the substrate stiffness in semilogarithmic scale, (B) the nuclear shape index versus the substrate stiffness in a semilogarithmic scale, and (C) the projected nuclear area as a function of the nuclear shape index. The red line corresponds to a linear fit ($R^2 = 0.997$). (For interpretation of the references to color in this figure legend, the reader is referred to the online version of this chapter.)

shape by introducing a nuclear shape index, such as: $NSI = (4 \times \pi \times \text{area})/\text{perimeter}^2$. We show that NSI decreased from ~0.95 to ~0.78 when HUVECs were plated on substrate of 2.5 kPa and 1 MPa, respectively, suggesting a large deformation of the nucleus on stiffer culture substrates (Fig. 3.5B). Interestingly, we demonstrated that the dependences of the nuclear area and NSI on the matrix stiffness are intimately linked, as observed by the linear relation in Fig. 3.5C. These results show that modifications of the cell matrix rigidity can significantly impact the nucleus without affecting other cellular parameters, such as shape, substrate adhesion, or spreading area.

GENERAL CONCLUSIONS

We have reported two microcontact printing methods on hydroxy-PAAm hydrogels and Sylgard 184 silicone elastomers for creating "soft" and "stiff" functionalized biomaterials, over three orders of magnitude. These methods, which are based on commercially available and low-cost polymers, do not require special equipment or specific expertize. We hope that these methods will permit biologists and biophysicists to tune easily the elastic moduli of biomaterials in order to decouple the effect of mechanotransduction cues such as matrix stiffness, cell shape, or spreading area on cell fate.

Acknowledgments

This work was supported by the Belgian National Foundation for Scientific Research (F.R.S.-FNRS) through "MIS Confocal Microscopy," "Crédit aux Chercheurs" grants, and the "Nanomotility FRFC project" (no. 2.4622.11). T.G. doctoral fellowship is supported by the Foundation for Training in Industrial and Agricultural Research (FRIA). M.R. is Research Fellow of the F.R.S.-FNRS.

References

Brock, A., Chang, E., Ho, C. C., LeDuc, P., Jiang, X., Whitesides, G. M., et al. (2003). Geometric determinants of directional cell motility revealed using microcontact printing. *Langmuir*, *19*, 1611–1617.

Chen, C. S., Mrksich, M., Huang, S., Whitesides, G. M., & Ingber, D. E. (1997). Geometric control of cell life and death. *Science*, *276*, 1425–1428.

Engler, A. J., Sen, S., Sweeney, H. L., & Discher, D. E. (2006). Matrix elasticity directs stem cell lineage specification. *Cell*, *126*, 677–689.

Gabriele, S., Benoliel, A.-M., Bongrand, P., & Théodoly, O. (2009). Microfluidic investigation reveals distinct roles for actin cytoskeleton and myosin II activity in capillary leukocyte trafficking. *Biophysical Journal*, *96*, 4308–4318.

Gabriele, S., Versaevel, M., Preira, P., & Théodoly, O. (2010). A simple microfluidic method to select, isolate, and manipulate single-cells in mechanical and biochemical assays. *Lab on a Chip*, *10*, 1459–1467.

Gao, L., McBeath, R., & Chen, C. S. (2010). Stem cell shape regulates a chondrogenic versus myogenic fate through Rac1 and N-cadherin. *Stem Cells*, *28*, 564–572.

Guilak, F., Cohen, D. M., Estes, B. T., Gimble, J. M., Liedtke, W., & Chen, C. S. (2009). Control of stem cell fate by physical interactions with the extracellular matrix. *Cell Stem Cell*, *5*, 17–26.

Grevesse, T., Versaevel, M., Circelli, G., Desprez, S., & Gabriele, S. (2013). A simple route to functionalize polyacrylamide hydrogels for the independent tuning of mechanotransduction Cues. *Lab on a Chip*, *13*, 777–780.

Hamieh, M., Al Akhrass, S., Hamieh, T., Damman, P., Gabriele, S., Vilmin, T., et al. (2007). Influence of substrate properties on the dewetting dynamics of viscoelastic polymer films. *Journal of Adhesion*, *83*, 367–381.

Jaalouk, D. E., & Lammerding, J. (2009). Mechanostransduction gone awry. *Nature Review Molecular Cell Biology, 10*, 63–73.

Kwon, M., Godinho, S. A., Chandhok, N. S., Ganem, N. J., Azioune, A., Théry, M., et al. (2008). Mechanisms to suppress multipolar divisions in cancer cells with extra centrosomes. *Genes Development, 22*, 2189–2203.

Lecuit, T., & Lenne, P. F. (2007). Cell surface mechanics and the control of cell shape, tissue patterns and morphogenesis. *Nature Review Molecular Cell Biology, 8*, 633–644.

Palchesko, N. P., Zhang, L., Sun, Y., & Feinberg, A. W. (2012). Development of polydimethylsiloxane substrates with tunable elastic modulus to study cell mechanobiology in muscle and nerve. *PLoS One, 7*, e51499.

Pathak, A., & Kumar, S. (2012). Independent regulation of tumor cell migration by matrix stiffness and confinement. *Proceedings of the National Academy of Sciences of the United States of America, 109*, 10334–10339.

Pelham, R. J., Jr., & Wang, Y. (1997). Cell locomotion and focal adhesions are regulated by substrate flexibility. *Proceedings of the National Academy of Sciences of the United States of America, 94*, 13661–13665.

Reiter, G., Al Akhrass, S., Hamieh, M., Damman, P., Gabriele, S., Vilmin, T., et al. (2009). Dewetting as an investigative tool for studying properties of thin polymer films. *European Physical Journal: Special Topics, 166*, 165–172.

Solon, J., Levental, I., Sengupta, K., Goerges, P. C., & Janmey, P. A. (2007). Fibroblast adaptation and stiffness matching to soft elastic substrates. *Biophysical Journal, 93*, 4453–4461.

Théry, M. (2010). Micropatterning as a tool to decipher cell morphogenesis and functions. *Journal of Cell Science, 123*, 4201–4213.

Théry, M., Racine, V., Pépin, A., Piel, M., Chen, Y., Sibarita, J. B., et al. (2005). The extracellular matrix guides the orientation of the cell division axis. *Nature Cell Biology, 7*, 947–953.

Thomas, C. H., Collier, J. H., Sfeir, C. S., & Healy, K. E. (2002). Engineering gene expression and protein synthesis by modulation of nuclear shape. *Proceedings of the National Academy of Sciences of the United States of America, 99*, 1972–1977.

Versaevel, M., Grevesse, T., & Gabriele, S. (2012). Spatial coordination between cell and nuclear shape within micropatterned endothelial cells. *Nature Commununications, 3*, 671.

Versaevel, M., Grevesse, T., Riaz, M., & Gabriele, S. (2013). Nuclear confinement: Putting the squeeze on the nucleus. *Soft Matter, 9*, 6665–6676.

Wang, N., Tytell, J. D., & Ingber, D. E. (2009). Mechanotransduction at a distance: Mechanically coupling the extracellular matrix with the nucleus. *Nature Review Molecular Cell Biology, 10*, 75–82.

Yeung, T., Georges, P. C., Flanagan, L. A., Marg, B., Ortiz, M., Funaki, M., et al. (2005). Effects of substrate stiffness on cell morphology, cytoskeletal structure, and adhesion. *Cell Motility and the Cytoskeleton, 60*, 24–34.

The Facile Generation of Two-Dimensional Stiffness Maps in Durotactic Cell Platforms Through Thickness Projections of Three-Dimensional Submerged Topography

C.H.R. Kuo*, J. Láng*,†, O. Láng†,‡, L. Kőhidai†, and E. Sivaniah*,‡

*Biological and Soft Systems, Cavendish Laboratory, University of Cambridge, Cambridge, United Kingdom

†Department of Genetics, Cell- and Immunobiology, Semmelweis University, Budapest, Hungary

‡Institute for Integrated Cell-Material Sciences, Kyoto University, Kyoto, Japan

CHAPTER OUTLINE

http://dx.doi.org/10.1016/B978-0-12-800281-0.00004-X

Abstract

An innovative platform that aims to facilitate studies of how adherent cells migrate in response to rigidity gradients or durotaxis has been developed. Soft polyacrylamide gel–based cell culture scaffolds are used to fabricate flat surfaces containing elasticity gradients through changes in the underlying patterned features. Moreover, this inert gel surface supports long-term cell viability and offers a tunable stiffness. By manipulating the thickness of the gel substrate through the embedded patterns, this system is also capable of directing collective cell patterning.

INTRODUCTION

In recent years, numerous mechanosensing studies have recognized the significance of mechanical cues on cellular response leading to changes in cellular structure and function similar to changes induced by chemical signals. Studies focusing on anchor-dependent cell behavior on surfaces with various elastic moduli reveal that finite mechanical differences could induce cell polarization and direct cell migration, a phenomenon defined as mechanotaxis (Lo, Wang, Dembo, & Wang, 2000). The effects may be further classified as tensotaxis when the movements in adherent cells can be guided by manipulating and creating tension on a flexible substrate. On the other hand, durotaxis is the effect during which the cell migrates preferentially along stiffness gradients on a compliant substrate. A range of cell behaviors including morphogenesis, cell-substratum adhesion, migration, proliferation, and differentiation are affected by the structure and mechanical properties of its microenvironment (Rehfeldt, Engler, Eckhardt, Ahmed, & Discher, 2007). Controlling such biological processes has become a long-standing goal in the development of functional biomaterials. The paucity of good generic testing platforms has posed major challenges in extracting relevant information required to design a functional artificial tissue matrix.

Despite continuous efforts to optimize in vitro testing platforms in mechanosensing studies so that cellular behavior mimics what is found in the in vivo tissue environment, there are still difficulties to overcome. First, the interpretation of results is often complicated by simultaneous changes in both chemical and physical parameters. In particular, studies have used compliant surfaces coated with protein-rich extracellular matrix (ECM) materials such as collagen; fibrin; or a mixture of collagen, laminin, and other proteins including bioactive growth factors forming Matrigel. These ECM protein coatings could form large-scale structures with feature sizes up to several hundred microns with variable local densities. The resulting topographical cues are known to cause a contact guidance effect where the cells tend to organize along the ECM fibers and use the ECM conduits for migration, leading to measurement bias (Sheetz, Felsenfeld, & Galbraith, 1998). In addition, cellular changes in genotypic traits are reported. The ECM proteins provide abundant integrin receptor binding cues, which are known to trigger an up-regulation of integrin binding subunits that are specific to the ECM (Martin, 1997). These chemical signals from the protein surfaces induce a cascade of complex biochemical signals that again result in measurement bias. The presence of ECM gradients could also direct haptotaxis, another mechanism for directing cell migration (Grinnell, 1986).

Second, the measurements are typically performed on a single migrating cell, as it is believed to help minimize potential cell-to-cell interactions. However, this does not accurately reflect the complex in vivo environment such as that in wound healing, where thousands of cells express collective migratory behavior and migrate toward a targeted area. Finally, most existing platforms are based on skill-demanding multigradient gel preparation or photolithography/nano-fabrication methodologies operating within expensive cleanroom facilities. These high skill and equipment requirements prevent most biological research laboratories from performing patterned rigidity mechanotaxis assays.

Motivated by these complications, a new, simple method has been developed to fabricate flat elastic substrates with a multirigidity landscape and to investigate cellular sensing mechanisms at single and multicellular levels (Kuo, Xian, Brenton, Franze, & Sivaniah, 2012). This work addresses some of the issues raised in the development of biologically inert, rigidity patterned polyacrylamide gel template for use as a model system to study the mechanosensitivity of tissue cells.

4.1 GLASS TEMPLATE TREATMENT

Here we present the method for pretreating the glass template and its top cover glass. The glass template treatment protocol is applied to treat various glass substrates to allow gel to bind strongly to the treated surface once it is cross-linked. The technique has been expanded to include several variations such as the step and the bead system (Fig. 4.1). A method is also presented for treating the top cover glass to allow for easy removal from the cross-linked gel. These treatments can be done in advance on large quantities of templates and top covers, and they can be stored for future usage.

4.1.1 Material

1. Glass coverslip 1, 24 mm × 24 mm (Academy)
2. Glass coverslip 2, 13 mm diameter No. 0 (Agar Scientific)
3. Grooved glass template, 25 mm × 25 mm borosilicate glass, 1.8 mm spacing between each groove, 0.2 mm × 0.2 mm × 25 mm groove cut (UQG Optics Ltd.)
4. Sodium hydroxide ≥98% pellets (NaOH) (Sigma Aldrich)

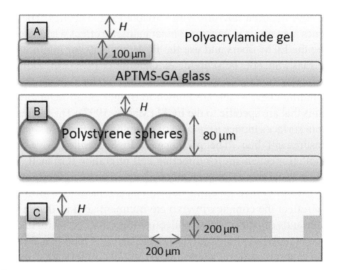

FIGURE 4.1

Illustration of various template cross-sections used to provide rigidity patterned scaffolds. From top to bottom: (A) Step template, (B) bead template, and (C) groove template. (For a color version of this figure, the reader is referred to the online version of this chapter.)

5. RainX Rain Repellent (Halfords)
6. (3-Aminopropyl)trimethoxysilane 97% (APTMS) (Sigma Aldrich)
7. Glutaraldehyde solution, 25 wt.% in H_2O (Sigma Aldrich)
8. UV-cured Adhesive 81 (Northland Products)
9. Duke Standards 4000 Series, 80 μm diameter monodispersed polystyrene spheres (Thermo Scientific)

4.1.2 Equipment

Fume cupboard (Schneider Elektronik)
UV lamp UVLS-28 EL series (UVP)
Ceramic hotplate (Fisher Scientific)

4.1.3 Method

1. To prepare the top glass coverslips, new coverslips, 13 and 24 mm are soaked in RainX solution for a few minutes, the excess is carefully wiped off with tissue paper, and the slide is dried and stored for future use.
2. To prepare the glass template, both the 13 mm coverslips and the 25 mm grooved glass surface are covered with 0.1 M NaOH solution for 1 min.
3. In a fume hood, remove the NaOH solution and add enough APTMS to cover the surface for 3 min.

4. Rinse the glass template surface with distilled water five times to ensure no APTMS remains. If any APTMS remains on the surface, in the subsequent step, it will react with the glutaraldehyde to form an orange precipitate.
5. After cleaning, dry the surface with a compressed air jet.
6. Cover the surface with 0.5% glutaraldehyde solution and allow to sit for 30 min.
7. Rinse the glass template with distilled water and dry the surface with compressed air.
   - The step template is prepared by gluing two 13-mm APTMS–glutaraldehyde-treated glass templates together using UV adhesives.
     a. Apply 1 µl of the UV adhesives to one edge of the 13-mm glass template.
     b. Gently put another 13-mm glass template on top of the adhesive so it covers roughly half of the bottom glass and forms a glass step.
     c. Put the resulting glass step into a UV box for 3 min to cure the adhesive.
   - The bead template is prepared by using monodispersed polystyrene spheres.
     a. Apply one drop of the polystyrene sphere solution on the center of the 13-mm glass template.
     b. Slowly air dry the liquid surface to enable the surface tension of the solvent to pull individual polystyrene spheres together as it evaporates away. Use tweezers to help guide the liquid drop to make sure only a densely packed monolayer array of beads remains in the end.
     c. Briefly treat the bead template with heat at 200 °C for 30 s to partially anneal the beads to the surface.
   - The groove template is prepared by custom preparation of slides. Typically, the blank slides are grooved using a diamond saw machining tool.

4.2 GEL PREPARATION

4.2.1 Material

1. Phosphate buffered saline (PBS) solution without Ca^{2+}/Mg^{2+} L1825 (Biochrom)
2. Acrylamide solution, 40% w/v (BDH)
3. N,N'-Methylene-bis-acrylamide solution, 2% w/v (Fisher Scientific)
4. Fluorescein O,O'-dimethacrylate (FDMA), 95% (Sigma Aldrich)
5. Dimethyl sulfoxide (DMSO) (Sigma Aldrich)
6. Ammonium persulphate (APS) (Sigma Aldrich)
7. N,N,N',N'-Tetramethylethylenediamine (TEMED) (Sigma Aldrich)
8. Hydrazine hydrate (Sigma Aldrich)
9. Acetic acid (Sigma Aldrich)
10. Poly-D-lysine hydrobromide (PDL) (Fisher Scientific)

4.2.2 Equipment

Fume cupboard (Schneider Elektronik)
Hera Safe KS12 Class II safety cabinet (Thermo Scientific)

Heraeus Fresco 17 centrifuge (Thermo Scientific)
Heraeus Heracell 240 incubator (Thermo Scientific)
Jencos grease-free desiccator (Fisher Scientific)
Pump FB70155 (Fisher Scientific)

4.2.3 Method

Polyacrylamide gel premixes of known bulk shear modulus were prepared according to the procedures developed by Moshayedi et al. (2010). The gel template is prepared on the day of cell deposition.

1. Use the table below to prepare a 500-µl gel premix of the desired mechanical properties. For gel staining, 1% (w/v) fluorescein FDMA in DMSO is added to the gel premix. In some cases, fluorescein may aggregate to form insoluble precipitate in this mixture. To remove the particulates, centrifuge at $10 \times g$ for 3 min and carefully transfer the supernatant to a new Eppendorf tube.

Shear modulus G′	100 Pa	300 Pa	1000 Pa	3.3 kPa	10 kPa	30 kPa
PBS (µl)	392.5	362.5	341	351	295	170
40% Acrylamide (µl)	62.5	62.5	94	94	150	225
2% Bis-Acrylamide (µl)	40	70	60	50	50	100
Stain (FDMA) (µl)	5	5	5	5	5	5

2. Place the gel premix in a vacuum desiccator for 15 min to remove all dissolved gas.
3. Add 1.5 µl TEMED and 5 µl 10% APS to the gel premix, mix the solution, and pipette 20–50 µl of the mixture onto the treated glass template. Avoid any intense mixing, which may introduce air into the solution. The amount of gel premix added will determine the final gel thickness.
4. Gently place the RainX-treated top glass coverslip. If the desired outcome is to produce gel thickness of less than 10 µm, place a small weight on top of the glass coverslip and use tissue paper to wipe away the excess gel premix.
5. Use the remaining gel premix in the tube to determine the extent of cross-link. Depending on the ratio of monomers and cross-linkers, complete cross-linking will take approximately 15 min.
6. Immerse the entire template in PBS for 30 min. Then gently slide the top coverslip off with the help of tweezers.
7. In the hood, treat the gel template with hydrazine hydrate for 3 h.
8. Remove the hydrazine hydrate, and treat the gel template with 5% (v/v) acetic acid in distilled water for 1 h.
9. Move the gel template into a sterile laminar flow cabinet, and wash the gel three times in sterile PBS for 10, 20, and 30 min, respectively.

10. To facilitate cell adhesion, the gels should be immersed in 100-μg/ml PDL solution and left in the incubator for at least 1 h.
11. Wash the gel three times in PBS using the same intervals as before.
12. Soak the gel in tissue culture media for at least 10 min prior to cell deposition.

4.3 CELL DEPOSITION

The method described below is for seeding 3T3 mouse fibroblasts on the gel template. However, the gel template has been successfully tested with the following cell lines: 3T3, HaCaT human keratinocyte, HepG2 human hepatocellular carcinoma, MCF-7 human breast carcinoma, and DLD1 human colorectal adenocarcinoma, and it could potentially be used for all adherent cell types. Nevertheless, cell types of different tissue origin exhibit varying bulk shear modulus preference and elastic moduli range to which they are able to comply and respond by durotaxis. 3T3 fibroblasts were described to durotact on substrates that had relatively wide ranges of mechanical properties (Kidoaki & Matsuda, 2008). On the other hand, durotaxis is also dependent on cellular motility that may result in kinetic differences between cell types exhibiting varying degrees of migratory capacity and may influence the optimal incubation time.

4.3.1 Material

1. 3T3 fibroblasts (ATCC: CCL-92)
2. PBS solution without Ca^{2+} / Mg^{2+} L1825 (Biochrom)
3. Dulbecco's Modified Eagle's Medium (DMEM) (Gibco)
4. Trypsin–Ethylenediaminetetraacetic acid (EDTA), 0.05% (Gibco)
5. Fetal bovine serum (Gibco)
6. Penicillin–streptomycin (PEN/STREP) (Gibco)

4.3.2 Equipment

Motic AE31 Trinocular inverted light microscope (Motic)
Hera Safe KS12 Class II safety cabinet (Thermo Scientific)
Heraeus Heracell 240 incubator (Thermo Scientific)
Centrifuge 5810R (Eppendorf)
Hemocytometer AC1000 (Hawksley)

4.3.3 Method

1. Prepare medium DMEM (450 ml), 10% FBS (50 ml), and 1% PEN/STREP (5 ml). Warm up in water bath for 15 min.
2. Remove flasks from the incubator, and examine the cultures under a microscope. Inspect the cell layer covering the flask surface. The minimal working density of

cultures is 90% confluency. If the flask is at least 90% confluent, the culture is ready for deposition.

3. In the hood, aspirate the media, and wash with PBS once.
4. Add 1 ml Trypsin–EDTA solution to the T-25 flask until the cell layer surface is covered. Store the flask inside the incubator for 5 min. Dislodge the cells from the flask surface by slapping the side of the flask with the heel of your hand 3–5 times. Examine under microscope to make sure all cells are detached.
5. Add 5-ml media to each T-25 Flask. Mix well with Trypsin–EDTA solution to neutralize. Collect fibroblast suspensions into a 15-ml polypropylene centrifuge tube, and centrifuge the contents at 800 rpm for 5 min.
6. Aspirate the spent media, and resuspend pellets in 1-ml fresh media.
7. Count the cell density of the fibroblast suspension using a hemocytometer, and dilute with fresh media to adjust the cell density to 50,000 cells/ml. If different cell types are used, cell density may be adjusted. A strong tendency to form aggregates may be overcome by lowering cell density by a factor of 2 and slightly increasing settling time. In contrast, using higher cell density may lead to difficulties during the image analysis as large cell aggregates and sheets may be formed, biasing image analysis.
8. Transfer individual gel templates to a new culture dish. For each 13-mm gel template, add 300 µl of the fibroblast suspension and 1 ml to each groove template. Allow the cells to settle for 30 min inside the incubator. In the case of cells exhibiting slow adhesion kinetics, settling time may be increased to 60 min.
9. Gently add 5 ml of fresh culture medium to each culture dish, and make sure all the templates are immersed. Store the dishes back in the incubator, and allow the cells to migrate for 16–24 h before imaging.

4.4 STAINING AND IMAGING

4.4.1 Material

1. Cell Tracker Orange CMRA (C34551, Invitrogen)
2. Syto 59 (red fluorescent, S11341, Invitrogen)
3. Leibovit's L15 medium (Sigma Aldrich)

4.4.2 Equipment

Zeiss LSM-510 confocal laser scanning microscope with a $25 \times$ objective (water immersion, NA = 0.8). It is equipped with Argon 488 nm, HeNe 543, 633 nm lasers with its corresponding BP 505-550, LP560, and LP650 filters.
Zeiss LSM Image Browser 4.2.0
ImageJ 1.47 (RSB)

4.4.3 **Method**

1. Add 5 μm of Syto 59 and 2.5 μm of Cell Tracker Orange to the gel templates, and incubate for 15 min.
2. Aspirate the old medium, and add in sufficient L15 medium to immerse the glass template. This medium allows the cell to remain viable during the imaging session without an environmental chamber.
3. Load the template onto the Zeiss confocal microscope, and start acquiring 3D image stack from random points on the gel. Each image stack will contain the transition zone between the shallow and deep regions of the gel. The z-stack is acquired from the surface of the gel to the surface of the glass template in a step size of 2 μm. This will allow an accurate determination of gel thickness in the subsequent analysis.
4. Analyze the image by counting the number of cells in the shallow N_s and deep N_d regions. The areas A of different regions are measured using the ImageJ software. The gel thickness H is measured from a 3D reconstruction of the z-stacks created using the Zeiss Image Browser.
5. The area normalized fraction of cells in the shallow region is calculated to quantify the level of preferential cell coverage. Cell distribution in the shallow region is calculated as $\phi_{c,s} = (N_s/A_s)/(N_s/A_s + N_d/A_d)$. The higher the value, the greater proportion of cells are found in the shallow region. When the cells remain homogeneously distributed across the whole area, $\phi_{c,s} \approx 50\%$. The average of multiple $\phi_{c,s}$ is plotted against its respective thickness H to determine the critical thickness at which cells are found to migrate.

4.5 **RESULTS**

The three templates discussed earlier were used to study durotaxis in 3T3 mouse fibroblasts (Kuo et al., 2012). When the fibroblasts were cultured on the step glass system, we found the durotaxis occurred with $H_{step} < 15$ μm (Fig. 4.2A). Furthermore, at below this critical depth, there existed an inverse relationship between H_{step} and $\phi_{c,s}$. The maximum amount of cells, between 75% and 90%, were located at the shallow region when the gel was thinnest; $H_{step} = 3$ μm (Fig. 4.2B). Surprisingly, changing the bulk shear modulus of the gel from $G' = 10$ to 3 kPa or 30 kPa does not affect the $\phi_{c,s}$ ($P > 0.05$ for $H_{step} \neq 7$ μm, Analysis of variance (ANOVA).

Similarly, we investigated the fibroblasts cultured on the bead template as a demonstration of the ability to easily incorporate complex alternating stiffness variations in the cell-culture surface. The complementary Atomic force microscopy (AFM) indentation study indicated that the slope of the apparent stiffness gradient was altered by the underlying patterns. The apparent stiffness gradient change was sharper over an abrupt step compared to the smoother gradient change over a bead. It was found that despite the smoother gradient, the critical gel thickness remained consistent, with $H_{bead} < 15$ μm (Fig. 4.2C). The inversely correlated association

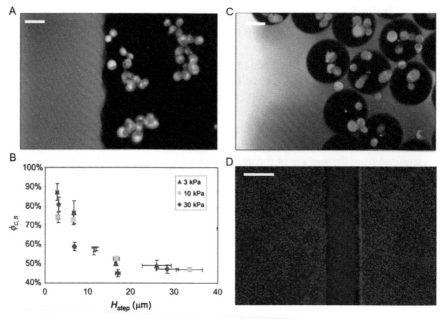

FIGURE 4.2

Cell response to complex mechanical substrates. (A) Confocal laser scanning microscopy (CLSM) images of fibroblasts growing on a step substrate where $H_{step}=3$ µm and $G'=10$ kPa. The cells are maximum intensity z-projections of the substrates corresponding to one optical plane. Fibroblasts relocate toward the apparently stiffer region of the substrate. (B) Plot of the area-normalized fraction of cells in the shallow region of the step substrate, $\phi_{c,s}$, as a function of gel thickness and bulk shear modulus G'. Within the investigated range, this critical thickness was largely independent of the bulk shear modulus G' of the hydrogel. Scale bar: 50 µm. (C) CLSM z-projection of fibroblasts grown on a bead substrate where $H_{bead}=8$ µm and $G'=10$ kPa. Cells migrated to the area over the beads where apparent stiffness is highest. Scale bar: 50 µm. (D) CLSM z-projection of fibroblasts grown on a bead substrate where $H_{groove}=6$ µm and $G'=10$ kPa. Cells homogeneously covered the surface of the stiffer region with no cells found over the softer groove region. Scale bar: 200 µm. (See color plate.)

Adapted with permission from Kuo et al. (2012).

between H_{bead} and $\phi_{c,s}$ was also unchanged, with approximately 80% of the cells being located on top of the beads for $H_{bead}=3$ µm. Again, this critical thickness was independent of the bulk shear modulus of the hydrogel ($P>0.05$ for all H_{bead}, ANOVA).

The groove template represents a slightly different proposition than a single step for the cells since the deep groove presents a finite area of apparently softer gel in between a much larger area of two shallower, stiffer gels to provide a different stiffness profile. Again, the critical gel thickness remained consistent, with

$H_{groove} < 15$ µm (Fig. 4.2D). This system demonstrated that durotaxis could have occurred even at high cell concentrations, creating the potential for massive mechanically guided cell-patterning scaffolds.

4.6 DISCUSSION

The rigidity patterned gel template provides an elegant solution by combining patterning methods and selective chemical modification schemes to control surface-matrix elasticity. It provides a controlled biocompatible surface that can be used to investigate durotactic interactions on the micrometer scales on which cells are organized. As demonstrated, three simple approaches to fabricate complex micrometer patterns provide promising tools to look into cell mechanosensitivity as well for the development of tissue engineering scaffolds. Moreover, the number of the possible diverse patterns to be created at the surface of the glass substrate and the number of the resulting gradient profiles are virtually unlimited. This allows a high flexibility in terms of both gradient shape (discontinuous or continuous) and strength, which in turn determine the range of biologically and medically relevant phenomena that can be studied or modeled in the described system. Significant differences were reported between stiffness gradients encountered by the cells *in vivo* in physiological or pathological conditions. Typically, physiological stiffness gradients occurring during tissue development were found to be shallower (<1 kPa/mm) than those characteristic for pathological events (>10 kPa/mm) such as myocardial infarcts or tumor formation (Tse & Engler, 2011).

It was demonstrated that with appropriate stiffness patterning, it is possible to control the placement of cells in an organized pattern on a chemically homogeneous substrate. Using the durotactic properties of fibroblasts, a strategy to measure the elastic interactions of active cells on rigidity patterned soft gel is proposed. By quantifying the average number of cells that have migrated to the stiffer side with varying gel depth, the level at which cells can feel through elastic tunable gels is discretely determined. In contrast with the previously developed techniques, this technique uses an unbiased charge-based PDL to promote cellular adhesion rather than the specific ECM protein-receptors based solutions. This allows for the unbiased study of the cellular signaling events underlying durotaxis that are also supposed to be dependent on matrix stiffness. This was shown for intracellular linker proteins at focal adhesion sites that underwent force-responsive unfolding to different extents depending on the stiffness of the substrate (Holle & Engler, 2011).

Finally, one significant achievement of this novel procedure is that the rigidity patterned surface requires neither advanced multigradient gel preparation nor access to cleanroom photolithographic patterning facilities. This can enable the study of cell migratory behavior in conventional biology laboratories and can contribute to the development of the next-generation scaffolds suitable for tissue engineering applications.

CONCLUSION

A better understanding of fibroblast migration is essential for the design of materials for regenerative medicine applications. Previous findings suggest that the substrate mechanics response is cell specific. Therefore, future work will apply this technique to study additional cell types involved in wound healing such as vascular endothelial cells and keratinocytes. Subsequently, only when cell-to-substrate and cell-to-cell interaction is understood as one integrated system will its application in tissue engineering material, pharmacology, and genetics be truly rational. As such, there remain many open questions regarding the physical background of mechanosensing, downstream signaling pathways, and the potential mechanisms of cell "memory." Our platform is a useful tool in the ongoing research aiming to answer these questions.

References

Grinnell, F. (1986). Focal adhesion sites and the removal of substratum-bound fibronectin. *The Journal of Cell Biology, 103*, 2697–2706.

Holle, A. W., & Engler, A. J. (2011). More than a feeling: discovering, understanding and influencing mechanosensing pathways. *Current Opinion in Biotechnology, 22*(5), 648–654.

Kidoaki, S., & Matsuda, T. (2008). Microelastic gradient gelatinous gels to induce cellular mechanotaxis. *Journal of Biotechnology, 133*, 225–230.

Kuo, C.-H. R., Xian, J., Brenton, J. D., Franze, K., & Sivaniah, E. (2012). Complex stiffness gradient substrates for studying mechanotactic cell migration. *Advanced Materials, 24*, 6059–6064.

Lo, C. M., Wang, H. B., Dembo, M., & Wang, Y. L. (2000). Cell movement is guided by the rigidity of the substrate. *Biophysical Journal, 79*, 144–152.

Martin, P. (1997). Wound healing—Aiming for perfect skin regeneration. *Science, 276*, 75–81.

Moshayedi, P., Costa, L. F., Christ, A., Lacour, S. P., Fawcett, J., Guck, J., et al. (2010). Mechanosensitivity of astrocytes on optimized polyacrylamide gels analyzed by quantitative morphometry. *Journal of Physics: Condensed Matter, 22*, 1–11.

Rehfeldt, F., Engler, A. J., Eckhardt, A., Ahmed, F., & Discher, D. E. (2007). Cell responses to the mechanochemical microenvironment—Implications for regenerative medicine and drug delivery. *Advanced Drug Delivery Reviews, 59*(13), 1329–1339.

Sheetz, M. P., Felsenfeld, D. P., & Galbraith, C. G. (1998). Cell migration: Regulation of force on extracellular-matrix-integrin complexes. *Trends in cell biology, 8*(2), 51–54.

Tse, J. R., & Engler, A. J. (2011). Stiffness gradient mimicking in vivo tissue variation regulate mesenchymal stem cell fate. *PLoS One, 6*(1), e15978.

CHAPTER

Micropatterning on Micropost Arrays

5

Nathan J. Sniadecki*,†, Sangyoon J. Han*, Lucas H. Ting*, and Shirin Feghhi*

*Department of Mechanical Engineering, University of Washington, Seattle, Washington, USA
†Department of Bioengineering, University of Washington, Seattle, Washington, USA

CHAPTER OUTLINE

ISSN 0091-679X
http://dx.doi.org/10.1016/B978-0-12-800281-0.00005-1

Abstract

Micropatterning of cells can be used in combination with microposts to control cell shape or cell-to-cell interaction while measuring cellular forces. The protocols in this chapter describe how to make SU8 masters for stamps and microposts, how to use soft lithography to replicate these structures in polydimethylsiloxane, and how to functionalize the surface of the microposts for cell attachment.

INTRODUCTION

The interplay between the mechanical properties of cells and the forces that they produce internally or applied to them externally play an important role in maintaining the normal function of cells. These forces also have a significant effect on the progression of mechanically related diseases. The tools and methods used to study the interplay between cell mechanics and cell forces have come from recent innovations in microfabrication and micropatterning.

Micropost arrays, also known as *microfabricated post array detectors* or micropillars, are arrays of vertical cantilevers, which are used to spatially measure the traction forces of cells attached to their tips via microscopy (Fig. 5.1). These tools have helped to elucidate the mechanical behavior of cells, the nature of cellular forces, and mechanotransduction (Bhadriraju et al., 2007; Cai et al., 2006; du Roure et al., 2005; Fu et al., 2010; Ganz et al., 2006; Ghibaudo et al., 2008; Grashoff et al., 2010; Han, Bielawski, Ting, Rodriguez, & Sniadecki, 2012; Lemmon, Chen, & Romer, 2009;

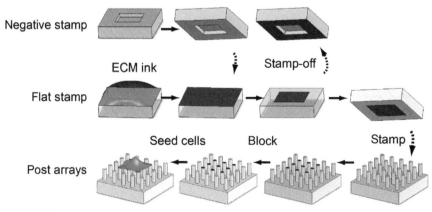

FIGURE 5.1

Stamping microposts for patterning cells. (See color plate.)

Lemmon et al., 2005; Liang, Han, Reems, Gao, & Sniadecki, 2010; Liu, Sniadecki, & Chen, 2010a; Liu, Tan, et al., 2010b; Nelson et al., 2005; Pirone et al., 2006; Rodriguez, Han, Regnier, & Sniadecki, 2011; Ruiz & Chen, 2008; Saez, Buguin, Silberzan, & Ladoux, 2005; Saez, Ghibaudo, Buguin, Silberzan, & Ladoux, 2007; Sniadecki et al., 2007; Sniadecki, Lamb, Liu, Chen, & Reich, 2008; Tan et al., 2003; Tee, Fu, Chen, & Janmey, 2011; Ting et al., 2012; Yang, Sniadecki, & Chen, 2007).

In this chapter, we describe the approach used to make the stamps and micropost arrays. Portions of the steps to make microposts have been described elsewhere and are useful sources of information (Desai, Yang, Sniadecki, Legant, & Chen, 2007; Sniadecki & Chen, 2007; Yang, Fu, Wang, Desai, & Chen, 2011). However, the techniques described here are more specific to the approach used recently in our lab (Han et al., 2012; Ting et al., 2012).

5.1 MICROFABRICATION OF SU-8 MASTERS FOR MICROPOSTS OR STAMPS

The following steps describe the process to fabricate an SU8 master for the micropost arrays or stamps. Two layers of SU8 are used for the features of the SU master for the micropost arrays. The first layer is used to form a strong adhesion with the silicon wafer. The second layer is used to create the features of the microposts or the stamps.

5.1.1 Materials

- Bare silicon wafer
- SU-8 2000 series photoresist (Microchem, Newton, MA). SU-8 2005 and 2010 are used in this example
- Disposable wide-mouth pipettes to dispense SU8 onto the silicon wafers
- SU-8 developer (Microchem, Newton, MA)
- Chrome mask with features for microposts or stamps
- Acetone
- Isopropanol alcohol (IPA)
- Deionized water (DI)
- Clean room wipes
- Fluorosilane (tridecafluoro-1,1,2,2-tetrahydrooctyl)-1-trichlorosilane) (T2492-KG, United Chemical Technologies).

5.1.2 Equipment

- Clean room environment (Class 1000 preferred)
- Photoresist spincoater with adjustable acceleration and velocity
- UV aligner
- Ozone cleaner (UVO cleaner; Jelight, Irvine, CA)

- Hotplates
- Clean Pyrex dishes
- Nitrogen gas
- Tweezers
- Digital timers.

5.1.3 Method

1. Clean the silicon wafer using a rinse of acetone for 10 s, a rinse of IPA for 10 s, and immersion in DI three times to remove organic residue that may be present on the wafer surface.
2. Place the wafer on a hotplate set to 200 °C for 10 min to dehydrate the wafer. This step is critical to ensure that the SU-8 adheres to the wafer surface. Cool the wafer to room temperature before the next step.
3. Spin a flat base layer of SU-8 2005 onto the wafer:
   a. Place the wafer on the spincoater and follow the manufacturer specifications for the chosen viscosity of SU-8 2000 series to set acceleration, rate, and time for spincoater. Some trial and error will be necessary to optimize the spinning process. Typically a base layer is thin, in the <5 μm thickness range so choose appropriately.
   b. Dispense the photoresist onto the wafer center in a smooth motion, using about 1 ml/in. in diameter of wafer to ensure even coverage. Allow the photoresist to spread momentarily before starting the spin cycle.
4. After the spin is completed remove the edge bead by either running an automatic edge bead removal cycle (available on most modern spincoaters) or by running a cleanroom wipe lightly wetted with SU-8 developer along the edge. This step is frequently necessary to ensure good quality photomask alignment in subsequent steps.
5. Place the wafer on a hotplate set to 95 °C for the manufacturer recommended preexposure bake time. A typical cycle is 3–5 min. This dries the SU-8 in place and prevents it from running.
6. Place the wafer onto a UV aligner. This step will not require a photomask because an unpatterned base layer is desired. Expose the wafer to UV light for the manufacturer recommended power and time. A typical time for the base photoresist is 3–10 s. An unpatterned base layer enables a strong adhesion for the second layer of SU-8 photoresist.
7. Place the wafer on a 95 °C hotplate for the manufacturer recommended postexposure bake time. A typical cycle is 3–5 min. Cool the wafer on a nonwoven wipe so that it does not cool too quickly and introduce thermal stress into the layer. If cracks in the SU8 film appear, they can be removed by annealing the wafer cyclically on a hotplate at 95 °C and cooling it several times.

8. Spin a second layer of SU-8 onto the wafer to create the micropost of the stamp pattern:
   a. Place the wafer on the spincoater and set the acceleration, rate, and time for the photoresist. A micropost layer will typically be 5–15 μm thick, while a stamp pattern will be from 5 to 50 μm thick.
   b. Repeat step 3b to dispense photoresist. A thicker photoresist will require more volume and more time will be required to allow it to spread properly.
9. Repeat step 4 to remove the edge bead.
10. Repeat step 5 for preexposure bake.
11. Move the wafer onto a UV aligner. Place the chrome mask into contact with the wafer ensuring that full contact is made between the mask and SU8 photoresist. Incomplete removal of edge beads can create a gap between the photoresist and the mask, which will cause diffraction of the UV light and poor pattern transfer. Expose the wafer to UV for the manufacturer recommended power and time. The micropost layer will typically require between 5 and 15 s, while a thicker stamp layer will require between 5 and 60 s.
12. Repeat step 7. Annealing may again be necessary to remove thermal stresses.
13. Develop the features by submerging the wafer in at least 1 cm deep dish of SU-8 developer. Agitate for the recommended time in developer. This will be between 5 and 15 min depending on SU-8 thickness, feature density, and feature size. Larger features with high aspect ratios or close packing will require more time for the developer to penetrate between the structures.
14. Remove the wafer, and while still wet, rinse with fresh SU-8 developer for 10 s on the top surface to remove the residual SU-8. Follow this rinse with an IPA rinse for 10 s. Use filtered nitrogen gas to dry the wafer.
15. A final annealing step may be used if cracks are seen in the SU8 structures. Place the wafer onto a 150 °C hotplate for 5 min and then cool slowly on a nonwoven wipe. This step has the potential to distort the features on the wafer due to reflow of the SU8 photoresist.
16. A final hard-bake step is recommended to fully cross-link the S-8 features on the master. A hard-baked master can be used for repeated polydimethylsiloxane (PDMS) casting without damage to the SU8 structures. Place the wafer onto a hot plate at 150 °C for 30 min, followed by a 200 °C hotplate for 30 min. Cool on a nonwoven wipe.
17. The wafer can now be cleaved and glued onto a glass slide to allow for easier handling and prevent breaking the master.
18. The master should be passivated using fluorosilane to prevent PDMS from permanently bonding its surface.
    a. Treat the master in an oxygen or air plasma for 60–120 s.
    b. Place the master into a silane desiccator with 50 μl of fluorosilane and place under vacuum. This vaporizes and binds the silane to the activated SU-8 surface; at least 1 h is recommended to achieve thorough coverage.

5.2 SOFT LITHOGRAPHY OF PDMS STAMPS AND MICROPOSTS

The following steps describe the process to fabricate PDMS stamps and micropost arrays from a SU8 master. PDMS stamps require a single-casting process from the SU8 master. Micropost arrays require a double-casting process from the SU8 master.

5.2.1 Materials

- SU8 masters for the stamp and micropost array
- Sylgard 184 (PDMS) base and curing agent (Dow Midland)
- Plastic cup
- Plastic stirrer (e.g., plastic 5 ml pipette)
- Aluminum foil, dish, or boat. Its size should match that of the SU8 master
- Razorblade
- Fluorosilane (tridecafluoro-1,1,2,2-tetrahydrooctyl)-1-trichlorosilane) (T2492-KG, United Chemical Technologies)
- Glass slide
- Glass pasture pipette.

5.2.2 Equipment

- Digital balance
- Vacuum desiccator chamber
- Convection oven
- Plasma chamber (Plasma Prep II, SPI, West Chester, PA)
- N_2 air.

5.2.3 Method for stamps

1. Prepare 20:1 mixture of base and curing agent by weight in a plastic cup using a digital scale.
2. Stir PDMS mixture for 5 min with a plastic stirrer.
3. Degas PDMS mixture in the desiccator for 1 h to remove air bubbles caused by stirring.
4. Place the SU8 master in an aluminum dish or boat and pour degassed PDMS over the master until it is 1 cm thick. Push down on the SU8 master at its corners using a disposable pipette to position it at the bottom of the aluminum dish.
5. Degas the PDMS for 30 min, or until almost all bubbles have dissipated. A gentle puff of N_2 gas can be used to pop air bubbles that are trapped at the surface.
6. Place the aluminum dish in a convection oven at 110 °C for 9–11 min or until the PDMS is firm.
7. Cut the aluminum dish away from the silicon master using a razorblade.

8. Cut away any access PDMS from the bottom of the SU8 master, and along the edges between the PDMS and the SU8 master.
9. Peel the PDMS away from the SU8 master very slowly.
10. Trim and cut PDMS to form 1 cm × 1 cm × 1 cm stamps. Cut a notch at one corner to mark the side opposite of the stamp features.

5.2.4 Method for negative molds for micropost

1. Prepare 10:1 mixture of base and curing agent by weight in a plastic cup using a digital scale.
2. Repeat steps 2–9 above for making PDMS stamps to create several negative copies of the SU8 master of the micropost arrays. These copies will be the negative molds used to make PDMS micropost arrays.
3. Plasma treat the negative molds inside the plasma chamber at 100 W for 90 s to activate the surface of the PDMS. Power and exposure times will differ if another plasma chamber is used.
4. Place negative molds into a glass desiccator with the features oriented face-up. Use tweezers to handle the negative molds.
5. Spread 25–50 μl of fluorosilane on a glass slide in the center of the desiccator using a glass pipette and place the pipette with fluorosilane in its tip on its side inside the desiccator.
6. Apply vacuum to the desiccator and expose the negative molds to fluorosilane vapor for 1 h.

5.2.5 Method for micropost arrays

1. Prepare 10:1 mixture of base and curing agent by weight in a plastic cup using a digital scale.
2. Stir PDMS mixture for 5 min with a plastic stirrer.
3. Degas PDMS mixture in the desiccator for 1 h to remove air bubbles caused by stirring.
4. Oxygen plasma treat the glass slides inside the plasma chamber at 100 W for 90 s.
5. Pipette the degassed PDMS into the negative molds so that the features are completely covered. Remove any air bubbles with a gentle puff of N_2 gas. If necessary, place structure in degasser to remove further bubbles.
6. Place the plasma-treated glass slides on top of the PDMS in the negative molds with the plasma-treated side of the glass facing the PDMS. Start with one edge of the glass on the negative mold and slowly lower it without trapping air bubbles between the PDMS and the slide.
7. Place the negative mold into a convention oven at 110 °C for 6 h in order for the PDMS to fully cure.
8. Remove the negative molds from the oven and allow them to cool to room temperature.

9. Slowly peel the glass slide away from the negative mold to release the micropost arrays on a glass slide. It is recommended to peel along the diagonal direction of the micropost array to prevent the microposts from sticking to each other.
10. Inspect the micropost array under a light microscope to determine whether the array has collapsed microposts, which is not useable.
11. Trim any excess PDMS that could interfere with stamping from the sides and edges.

5.2.6 Discussion of micropost arrays

If we wish to use the same negative mold again, we should treat the molds with plasma and fluorosilane every two castings. However, we will need to inspect the mold under the light microscope before each use to ensure that it has not developed any surface cracks for distortions in the array (Tooley, Feghhi, Han, Wang, & Sniadecki, 2011).

To watch cells on microposts live under an inverted microscope, a glass coverslip should be used instead of a glass slide. The thickness from the top surface of microposts to the objective lens should be less than the working distance of an objective. For example, the working distance for a $40 \times$ oil immersion objective is approximately 200 μm. Thus, we use the microposts (on a cover glass) to the bottom surface of a Petri dish with a hole directly.

5.3 STAMPING MICROPOST ARRAYS FOR CELL CULTURE

A stamp-off method is used to print a pattern of extracellular matrix onto the tips of the microposts. These steps are minor modification of the steps described in Chapter 1. Two types of stamps are made: a "flat stamp" made by casting against a featureless master and a "negative stamp" with features used for removing protein from the flat stamps, that is, negative features (Fig. 5.1).

5.3.1 Materials

- PDMS flat stamp (1:20 PDMS)
- PDMS negative stamp (1:20 PDMS)
- PDMS micropost arrays (1:10 PDMS)
- Sterile DI
- 50 μg/ml human fibronectin (FN) in sterile DI water (BD Biosciences)
- Ethanol (100% and 70%)
- 5 μg/ml bovine serum albumin (BSA) conjugated with Alexa Fluor 594 (A13101, Invitrogen) or 1 μg/ml fluorescent lipophilic carbocyanine (DiI) solution prepared in DI (D3886, Invitrogen)
- 0.2% Pluronic F127 (BASF, Mount Olive, NJ)
- Phosphate buffer solution (PBS)
- Aluminum foil
- 100-mm Petri dish.

5.3.2 **Equipment**
- Sterile biosafety cabinet
- UV-ozone (UVO cleaner; Jelight, Irvine, CA)
- Nitrogen gas
- Sterile tweezers.

5.3.3 **Method**
1. All steps are to be completed while working in a sterile biosafety cabinet.
2. Place flat stamps in a Petri dish with working side up. Place droplets of FN solution onto the corners of the stamps, then on the edges, and lastly on the interior regions. As the protein in solution is adsorbed on the PDMS surface, the surface of the stamp becomes hydrophilic. This process of starting at the corners and edges requires a lower volume of FN solution to coat the stamps.
3. Incubate each stamp with FN solution for 1 h.
4. Wash stamps with DI water by pouring it into the dish at the edge and letting the water level rise over the stamps.
5. Transfer stamps into a second Petri dish filled with DI water. After washing, dry them with N_2.
6. Treat the negative stamps with UV-ozone for 7 min to make them more hydrophilic. This step is performed outside of the biosafety cabinet.
7. Using tweezers, place the patterned side of the negative stamps in contact with the protein-coated side of the flat stamps. Tap lightly on the stamps with tweezers to ensure full contact. Remove the stamps 5 s after contact. The flat stamps are now ready for patterning the microposts.
8. Treat the micropost arrays with UV-ozone for 7 min to activate the micropost surface for stamping. This step is performed outside of the biosafety cabinet.
9. Place the flat stamp in contact with the microposts. Tap lightly the top of the stamp with tweezers to ensure all areas have good contact, but be careful not to collapse the posts. It is helpful to practice this step under an inverted light microscope to figure out how much tapping pressure is needed. Carefully remove flat stamps with tweezers.
10. Submerge the micropost arrays in a Petri dish with 100% ethanol for 10 s. This step is used to wet the surface and then subsequent dilution washes the ethanol away.
11. Transfer micropost arrays to a Petri dish with 70% ethanol for 10 s. This step has a secondary benefit of sterilizing the micropost substrates and so from this point forward substrates should be handled using aseptic techniques.
12. Transfer micropost arrays to a Petri dish with sterile DI water for 10 s. The PDMS microposts are very hydrophobic and the surface tension of water can cause them to collapse together. Take caution in keeping the arrays from dewetting during transfer by moving carefully between dishes.
13. Repeat step 12 with two more dishes of sterile DI water.

14. Transfer the micropost arrays to a Petri dish with the fluorescent dye.
    a. For BSA, place droplets of BSA solution onto the bottom surface of a Petri dish and place the micropost arrays upside down onto each droplet.
    b. For DiI, submerge the micropost arrays in a Petri dish with DiI solution. DiI provides a stronger fluorescent signal than BSA.
15. Cover with aluminum foil and incubate with dye for 1 h.
16. Wash microposts with DI water for 10 s. For DiI, rinse the microposts thoroughly with sterile DI water after treatment to remove excess DiI, which can be taken up into the membranes of the cells during cell culture.
17. Place droplets of Pluronic solution in a Petri dish and place the micropost arrays upside down onto each droplet. Incubate for 30 min. This step is used to block the unstamped portions of the micropost, including the base and sidewalls, so that cells will not adhere there.
18. Repeat step 12 with two dishes of sterile DI water. Pluronic treatment makes the surface of PDMS hydrophilic so less caution is needed in transferring between dishes since they are not easily dewetted.
19. Transfer microposts into a dish with PBS. Microposts arrays in PBS can be stored at 4 °C for up to 1 week, but better results are obtained if used immediately for cell plating.

5.3.4 Discussion

Stamping is an essential step in the biofunctionalization of micropost arrays. During this step the matrix of interest is transferred to the tips of the posts for cells to adhere and spread on. We have been able to stamp fibronectin, fibrinogen, von Willebrand factor, collagen type I, and collagen type IV.

Stamping is challenging on softer microposts (low spring constant) for a large number of them collapse under the pressure of the stamp or during placement or removal of the stamp. In these cases occasionally stamping protocols are modified by stamping substrates under water. However, this modification adds to the limitations of the preparation of the substrate since once the posts are wet all the rest of protocol should also be done in a wet environment to avoid collapsing the micropost due to the force of water's surface tension.

An alternative method can be used to adsorb the protein of interest to the tips of the microposts which bypasses some of the limitations discussed. This method is called the *droplet adsorption method* and is based on the interaction of a water-based droplet with a hydrophobic surface. Micropost arrays are made of PDMS which presents hydrophobic characteristics and the topology of the microposts also adds to the overall hydrophobicity of the surface. In this method the protein of interest is dissolved in a water-based solution. The droplet is then placed on the micropost array. The overall hydrophobicity of the substrate prevents the droplet from penetrating between the microposts and wetting the side walls of the microposts. Therefore, the protein is only transferred to the tips of the posts, which is equivalent to regular stamping for stiffer substrates.

5.4 SEEDING CELLS

Follow these general procedures to culture cells onto the micropost arrays. Cell detachment procedures may differ depending on the cell type.

5.4.1 Materials

- Chosen cell line
- Trypsin–Ethylenediaminetetraacetic acid (EDTA)
- Growth medium
- Tissue culture dish
- PBS.

5.4.2 Equipment

- Sterile biosafety cabinet
- Sterile tweezers.

5.4.3 Method

1. Cells should be at or near confluence in the starting tissue culture dish or flask. This is necessary so that a sufficient number of cells are available to land on the microposts.
2. Micropost substrates should be placed onto a tissue culture dish containing prewarmed media and allowed to incubate for 1 h prior to seeding.
3. Aspirate away the growth media in the cell culture dish or flask.
4. Rinse the cells with PBS to remove residual media in the dish.
5. Detach the cells from the dish or flask using trypsin–EDTA solution. A typical volume is 1 ml of trypsin–EDTA per 50 cm^2 of surface area; however, this will depend on the cell type and cell age. Cells that have been cultured longer will typically require more time. Place the cells with the trypsin–EDTA solution in an incubator and check periodically to see if they have detached from the surface.
6. Once the cells have detached, resuspend the cells in prewarmed media with gentle agitation to separate groups into individual cells.
7. Pipette the cell solution into the dish containing the micropost substrates. Place this dish into the incubator and allow cells to settle onto the microposts. Take periodic observations to gauge the density of cells that have landed on the microposts. If a significant portion of them have settled, then a rinse with fresh prewarmed media can be done to remove unattached cells. This step may or may not be necessary depending on the type of cell being seeded.
8. Allow at least 24 h for the cells to spread and stabilize on the micropost tips before traction force measurements are taken.

GENERAL CONCLUSIONS

The steps described in this chapter cover the fabrication of stamps and microposts. These techniques are meant as a starting point for using microposts and not the *de facto* approach. We have found it necessary to adjust the steps or conditions in order to optimize the process to suit a particular set of experiments or cell type. Furthermore, it is highly recommend that a new user should watch a visual demonstration of some of the techniques used for micropost arrays in order to gain an understanding of how to work with them during their preparation (Desai et al., 2007).

References

Bhadriraju, K., Yang, M., Alom Ruiz, S., Pirone, D., Tan, J., & Chen, C. S. (2007). Activation of ROCK by RhoA is regulated by cell adhesion, shape, and cytoskeletal tension. *Experimental Cell Research, 313*(16), 3616–3623.

Cai, Y., Biais, N., Giannone, G., Tanase, M., Jiang, G., Hofman, J. M., et al. (2006). Nonmuscle myosin IIA-dependent force inhibits cell spreading and drives F-actin flow. *Biophysics Journal, 91*(10), 3907–3920.

Desai, R. A., Yang, M. T., Sniadecki, N. J., Legant, W. R., & Chen, C. S. (2007). Microfabricated post-array-detectors (mPADs): An approach to isolate mechanical force. *Journal of Visualized Experiments*, (7), 311.

du Roure, O., Saez, A., Buguin, A., Austin, R. H., Chavrier, P., Siberzan, P., et al. (2005). Force mapping in epithelial cell migration. *Proceedings of the National Academy of Sciences of the United States of America, 102*(7), 2390–2395.

Fu, J., Wang, Y. K., Yang, M. T., Desai, R. A., Yu, X., Liu, Z., et al. (2010). Mechanical regulation of cell function with geometrically modulated elastomeric substrates. *Nature Methods, 7*(9), 733–736.

Ganz, A., Lambert, M., Saez, A., Silberzan, P., Buguin, A., Mege, R. M., et al. (2006). Traction forces exerted through N-cadherin contacts. *Biology of the Cell, 98*(12), 721–730.

Ghibaudo, M., Saez, A., Trichet, L., Xayaphoummine, A., Browaeys, J., Silberzan, P., et al. (2008). Traction forces and rigidity sensing regulate cell functions. *Soft Matter, 4*(9), 1836–1843.

Grashoff, C., Hoffman, B. D., Brenner, M. D., Zhou, R., Parsons, M., Yang, M. T., et al. (2010). Measuring mechanical tension across vinculin reveals regulation of focal adhesion dynamics. *Nature, 466*(7303), 263–266.

Han, S. J., Bielawski, K. S., Ting, L. H., Rodriguez, M. L., & Sniadecki, N. J. (2012). Decoupling substrate stiffness, spread area, and micropost density: A close spatial relationship between traction forces and focal adhesions. *Biophysics Journal, 103*(4), 640–648.

Lemmon, C. A., Chen, C. S., & Romer, L. H. (2009). Cell traction forces direct fibronectin matrix assembly. *Biophysics Journal, 96*(2), 729–738.

Lemmon, C. A., Sniadecki, N. J., Ruiz, S. A., Tan, J. L., Romer, L. H., & Chen, C. S. (2005). Shear force at the cell–matrix interface: Enhanced analysis for microfabricated post array detectors. *Mechanics and Chemistry of Biosystems, 2*(1), 1–16.

Liang, X. M., Han, S. J., Reems, J. A., Gao, D., & Sniadecki, N. J. (2010). Platelet retraction force measurements using flexible post force sensors. *Lab on a Chip, 10*(8), 991–998.

Liu, Z., Sniadecki, N. J., & Chen, C. S. (2010a). Mechanical forces in endothelial cells during firm adhesion and early transmigration of human monocytes. *Cellular and Molecular Bioengineering, 3*(1), 50–59.

Liu, Z., Tan, J. L., Cohen, D. M., Yang, M. T., Sniadecki, N. J., Ruiz, S. A., et al. (2010b). Mechanical tugging force regulates the size of cell–cell junctions. *Proceedings of the National Academy of Sciences of the United States of America, 107*(22), 9944–9949.

Nelson, C. M., Jean, R. P., Tan, J. L., Liu, W. F., Sniadecki, N. J., Spector, A. A., et al. (2005). Emergent patterns of growth controlled by multicellular form and mechanics. *Proceedings of the National Academy of Sciences of the United States of America, 102*(33), 11594–11599.

Pirone, D. M., Liu, W. F., Ruiz, S. A., Gao, L., Raghavan, S., Lemmon, C. A., et al. (2006). An inhibitory role for FAK in regulating proliferation: A link between limited adhesion and RhoA-ROCK signaling. *Journal of Cell Biology, 174*(2), 277–288.

Rodriguez, A. G., Han, S. J., Regnier, M., & Sniadecki, N. J. (2011). Substrate stiffness increases twitch power of neonatal cardiomyocytes in correlation with changes in myofibril structure and intracellular calcium. *Biophysics Journal, 101*(10), 2455–2464.

Ruiz, S. A., & Chen, C. S. (2008). Emergence of patterned stem cell differentiation within multicellular structures. *Stem Cells, 26*(11), 2921–2927.

Saez, A., Buguin, A., Silberzan, P., & Ladoux, B. (2005). Is the mechanical activity of epithelial cells controlled by deformations or forces? *Biophysics Journal, 89*(6), L52–L54.

Saez, A., Ghibaudo, M., Buguin, A., Silberzan, P., & Ladoux, B. (2007). Rigidity-driven growth and migration of epithelial cells on microstructured anisotropic substrates. *Proceedings of the National Academy of Sciences of the United States of America, 104*(20), 8281–8286.

Sniadecki, N. J., Anguelouch, A., Yang, M. T., Lamb, C. M., Liu, Z., Kirschner, S. B., et al. (2007). Magnetic microposts as an approach to apply forces to living cells. *Proceedings of the National Academy of Sciences of the United States of America, 104*(37), 14553–14558.

Sniadecki, N. J., & Chen, C. S. (2007). Microfabricated silicone elastomeric post arrays for measuring traction forces of adherent cells. *Methods in Cell Biology, 83*, 313–328.

Sniadecki, N. J., Lamb, C. M., Liu, Y., Chen, C. S., & Reich, D. H. (2008). Magnetic microposts for mechanical stimulation of biological cells: Fabrication, characterization, and analysis. *Review of Scientific Instruments, 79*(4), 044302.

Tan, J. L., Tien, J., Pirone, D. M., Gray, D. S., Bhadriraju, K., & Chen, C. S. (2003). Cells lying on a bed of microneedles: An approach to isolate mechanical force. *Proceedings of the National Academy of Sciences of the United States America, 100*(4), 1484–1489.

Tee, S. Y., Fu, J., Chen, C. S., & Janmey, P. A. (2011). Cell shape and substrate rigidity both regulate cell stiffness. *Biophysics Journal, 100*(5), L25–L27.

Ting, L. H., Jahn, J. R., Jung, J. I., Shuman, B. R., Feghhi, S., Han, S. J., et al. (2012). Flow mechanotransduction regulates traction forces, intercellular forces, and adherens junctions. *American Journal of Physiology—Heart and Circulatory Physiology, 302*(11), H2220–H2229.

Tooley, W. W., Feghhi, S., Han, S. J., Wang, J., & Sniadecki, N. J. (2011). Fracture of oxidized polydimethylsiloxane during soft lithography of nanopost arrays. *Journal of Micromechanics and Microengineering, 21*(5), 054013.

Yang, M. T., Fu, J., Wang, Y. K., Desai, R. A., & Chen, C. S. (2011). Assaying stem cell mechanobiology on microfabricated elastomeric substrates with geometrically modulated rigidity. *Nature Protocols, 6*(2), 187–213.

Yang, M. T., Sniadecki, N. J., & Chen, C. S. (2007). Geometric considerations of micro- to nanoscale elastomeric post arrays to study cellular traction forces. *Advanced Materials, 19*(20), 3119–3123.

Development of Micropatterned Cell-Sensing Surfaces

Jungmok You, Dong-Sik Shin, and Alexander Revzin

Department of Biomedical Engineering, University of California, Davis, California, USA

CHAPTER OUTLINE

Abstract

Microfabricated surfaces have been widely utilized for defining adhesion of single cells or groups of cells of various kinds. Beyond simple control of cell attachment, it is often important to monitor the molecules released by cells. Co-immobilizing miniature sensors alongside cells enables more sensitive detection of secreted factors and may allow for such detection to happen within the context of local microenvironment. Methods for interfacing cells and sensors are central to the notion of local *in situ* detection of cell function. This chapter describes the use of hydrogel photolithography for integrating cells and sensing elements on culture surfaces. Two types of micropatterned sensing surfaces are described: (1) arrays of microwells for single cell capture that contain antibodies against secreted proteins and (2) entrapment of enzymes inside hydrogel microstructures for local detection of cell metabolism. In both cases, poly(ethylene glycol) hydrogel lithography was employed to control cell attachment, in the second approach hydrogel structures also carried enzymes and functioned as sensors. The development of robust cell/sensor interfaces has implications for diagnostics, tissue engineering, and drug screening.

INTRODUCTION

Defining cellular interactions and detecting cell function on micropatterned surfaces have relevance to cellular/tissue engineering, diagnostics, and drug screening. There are a number of surface micropatterning approaches, including microarraying, microcontact printing, and photolithography that have been used to define composition of the local cellular microenvironment (Bhatia, Yarmush, & Toner, 1997; Co, Wang, & Ho, 2005; Khademhosseini, Langer, Borenstein, & Vacanti, 2006; Singhvi et al., 1994; Takayama et al., 1999; Whitesides, Ostuni, Takayama, Jiang, & Ingber, 2001). However, methods for analysis of cellular micropatterns remain limited to either immunofluorescent staining or bulk measurements by enzyme-linked immunosorbent assays or polymerase chain reactions. These bulk analysis techniques do not have the ability to resolve function of specific groups of cells on micropatterned surfaces. Despite the development of microfabricated sensing devices (Borgmann et al., 2006; Cai et al., 2002; Jung, Gorski, Aspinwall, Kauri, & Kennedy, 1999; Park & Shuler, 2003; Troyer, Heien, Venton, & Wightman, 2002; Voldman, Gray, & Schmidt, 1999), the integration of cells and sensors remains a significant challenge.

In this chapter we describe a strategy for seamless integration of cells and biosensors on micropatterned surfaces (Lee et al., 2009; Yan, Sun, Zhu, Marcu, & Revzin, 2009; Zhu et al., 2009). This strategy leverages the use of poly(ethylene glycol) (PEG) hydrogels that have the following advantages: (1) miniaturization with photolithography-like approaches, (2) nonfouling properties to allow control over cellular attachment (Revzin et al., 2004; Revzin, Sekine, Sin, Tompkins, & Toner, 2005; Revzin, Tompkins, & Toner, 2003; Suh, Seong, Khademhosseini, Laibinis, & Langer, 2004), and (3) hydrogels are an excellent matrix for entrapment of biomolecules such as enzymes (Allcock, Phelps, Barrett, Pishko, & Koh, 2006; Revzin et al., 2001; Russell & Pishko, 1999). In this approach, biorecognition elements such as enzyme and antibodies (Abs) may either be immobilized on surfaces

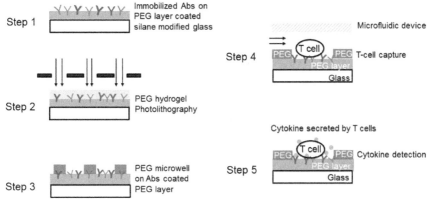

FIGURE 6.1

Fabrication of hydrogel microwells for capturing single T-cells and secreted cytokines. Step 1: Anti-CD4 and anti-IFN-γ Abs are physiadsorbed onto a lyophilized, PEG hydrogel-coated glass slide. Step 2: Photosensitive PEG prepolymer spincoated on top of the Ab-containing hydrogel layer was exposed to UV through a photomask. Step 3: Prepolymer not exposed to UV is removed by development in water, leaving behind microwells with nonfouling walls and Ab-containing attachment sites. Step 4: PEG hydrogel microwells are enclosed inside a PDMS microfluidic device and are incubated with RBC depleted human blood. Step 5: Mitogenic activation induces cytokine production in T-cells. Secreted cytokines become captured in the proximity of cells and are detected using sandwich immunoassay. (See color plate.)

underneath the cells (Fig. 6.1) or encapsulated inside PEG hydrogels microstructures next to the cells (Fig. 6.3B).

We envision cell culture systems with integrated sensors opening up new ways to concurrently define and detect cellular function.

6.1 MATERIALS

1. Glass slides (75 mm × 25 mm) (VWR International)
2. (3-Acryloxypropyl)trichlorosilane (Gelest, Catalog # SIA0199.0)
3. AZ 5214-E positive photoresist and AZ 300 MIF developer solution (Mays Chemical)
4. Sulfuric acid, hydrogen peroxide, ethanol, collagen from rat tail (type I, Catalog # C7661), epidermal growth factor (Catalog # E4127), horseradish peroxidase (HRP), poly(ethylene glycol)-diacrylate (PEG-DA) (MW 575, Catalog # 437441), 2,2′-dimethoxy-2-phenylacetophenone (DMPA) (Catalog #196118), 2-hydroxy-2-methylpropiophenone (Catalog #405655), paraformaldehyde (Catalog #P6148), TWEEN 20 (Catalog #P9416), rabbit antimouse IgG

antibody (Catalog #M7023), Na$_4$EDTA, KHCO$_3$, NH$_4$Cl, anhydrous toluene (99.9%) (Catalog #244511), bovine serum albumin (BSA), phorbol 12-myristate 13-acetate (PMA), and ionomycin (Sigma-Aldrich).

5. Dimethyl sulfoxide (DMSO) (Pierce)
6. 10× phosphate-buffered saline (PBS) (Cambrex)
7. Dulbecco's modified Eagle's medium (DMEM), minimal essential medium, sodium pyruvate, nonessential amino acids, L-glutamine, fetal bovine serum (FBS), Amplex red reagent (Catalog # A12222), and collagen I-fluorescein isothiocyanate (FITC) (Invitrogen Life Technologies)
8. Cell culture medium RPMI 1640 with L-glutamine (VWR)
9. Sylgard 184 Silicone Elastomer for poly-(dimethylsiloxane) (PDMS) channel preparation (Dow Corning)
10. Glucagon and insulin (Eli Lilly)
11. Hydrocortisone (Pfizer)
12. FluoroLink Cy3 reactive dye (Amersham Bioscience, Catalog # PA23500)
13. CellTracker green CMFDA and red CMTPX probes for long-term tracing of living cells (Molecular Probes, Catalog # C-2925 and C-34552)
14. Mouse antihuman CD4 Abs (13B8.2) (Beckman-Coulter, Catalog #IM0398)
15. Mouse antihuman IFN-γ Ab (clone K3.53) (Catalog #MAB2852), human recombinant IFN-γ (Catalog #285-IF-100), and biotinylated goat antihuman IFN-γ Ab (Catalog #BAF285) (R&D Systems)
16. Mouse IgG2a (OX34) (Serotec Antibodies, Catalog #MCA929)
17. anti-CD4-PE(L120) (BD Pharmingen, Catalog #340419)
18. FITC-conjugated Avidin (Pierce, Catalog # 21221).

6.2 METHODS

In this chapter, we discuss two methods for creating micropatterned cell sensing surfaces. Section 6.2.1 describes fabricating of PEG hydrogel microwells on glass containing cell capture and cytokine detection Abs (Fig. 6.1) (Zhu et al., 2009). In this case, a mixture of anti-CD4 and anti-IFN-γ Abs was physisorbed on a lyophilized, PEG hydrogel-coated glass slide. Subsequently, PEG hydrogel microwells of 20 μm diameter were fabricated on top of the Ab-containing hydrogel layer and were used for capture of CD4 T-cells and detection of secreted IFN-γ. Section 6.2.2 describes the method for integrating sensing hydrogel microstructures with micropatterned co-cultures (Fig. 6.3B) (Lee et al., 2009).

6.2.1 Integration of sensing Abs with single T-cells cultures

6.2.1.1 Silane treatment of glass substrates

This method starts with the modification of glass slides with silane coupling agent to ensure covalent attachment of PEG hydrogel films to glass slides (Fig. 6.1). The glass slides were cleaned by immersion in piranha solution consisting of a 1:1 ratio of 95% v/v sulfuric acid and 35% w/v of hydrogen peroxide for 10 min. Caution is required upon handling piranha solution. Glass slides were thoroughly rinsed with deionized

(DI) water and dried under nitrogen. Prior to silane modification, the glass slides were treated in an oxygen plasma chamber (YES-R3, San Jose, CA) at 300 W for 5 min. This process cleans the surface and creates reactive oxygen residues that can react with chlorosilane groups. For silane modification, oxygen plasma-treated glass slides were placed in a solution containing 3-acryloxypropyl trichlorosilane diluted in anhydrous toluene (final concentration = 20 mM, e.g., 20 μl/40 ml) for 10 min. The reaction was performed in a glove bag (AtmoBag, Sigma-Aldrich) filled with nitrogen to eliminate atmospheric moisture. The slides were rinsed with fresh toluene, dried under nitrogen, and cured at 100 °C for 2 h. The silane-modified glass slides can be stored in a desiccator prior to use. The deposition of silane layer can be confirmed by contact angle measurement. Typically contact angle of a silane modified glass slide is 55–60 °C.

6.2.1.2 Immobilization of cell- and cytokine-specific Abs onto PEG hydrogel-coated glass slides

PEG hydrogel films on glass slides were fabricated from the precursor solution of PEG-DA (MW 575) with 1% w/v photoinitiator, DMPA. This solution was spun at 700–1000 rpm for 5 s onto the silane-treated glass surface using a spin-coater (Spintech S-100, Redding, CA). The uniform layer of the PEG-DA precursor solution on glass was then exposed to UV light (365 nm, 60 mW/cm^2) from a light source (Omnicure 1000 light source, EXFO, Mississauga, Ontario, Canada). The UV exposure time ranging from 0.5 to 1 s caused free-radical polymerization and cross-linking of the diacrylated PEG. In addition to cross-linking, acrylated group in PEG molecules covalently bind to the acrylate groups on the glass substrate.

The PEG gel-coated glass slides were dehydrated by lyophilization for 48 h to ensure the rapid adsorption and uniform distribution of Abs upon printing. These dehydrated slides can be stored in a dessicator prior to use. For Abs printing, a mixture of purified anti-CD4 and IFN-γ Abs was dissolved in 1 × PBS at a concentration of 0.12 and 0.2 mg/ml and supplemented with Tween 20 (0.005% v/v). This Ab cocktail solution was manually pipetted onto the PEG gel-coated surface to create three Ab spots (∼1 mm, 0.5 mm print volume) in a row. Alternatively, a manual arrayer (Micro-Caster, Schleicher & Schuell, Keene, NH, Catalog # 10485047) containing eight pins was used to dispense small print volumes (20–70 nl per spot) of the Ab cocktail solution onto the PEG surface to form a 2 × 5 array of Ab spots (∼500 μm). Protein and Tween 20 concentrations need to be adjusted to control the quality of spot. In general, too higher concentration of the Tween 20 resulted in irregular shape of spot. After printing, surfaces were air-dried and stored at 4 °C prior to further use. Immunofluorescent labeling was employed routinely to characterize morphology and quality of printed Ab spots.

6.2.1.3 Fabrication of PEG gel microwells on top of Abs-coated hydrogel layer

The Abs immobilized on a PEG gel-coated glass substrate was covered with a pre-polymer comprised of PEG-DA and 2% (v/v) 2-hydroxy-2- methyl-propiophenone and spin-coated at 950 rpm for 4 s. Alternatively, a coverslip can be used to form a thin layer of PEG precursor solution. In this case, small amount of PEG precursor solution (5–10 µl per 20 × 24 mm glass piece) was added to the glass piece and then sandwiched between another glass substrate coverslip (~22 mm × 22 mm). The dimensions were made different for ease of separation of coverslips after UV exposure. When working with spin-coated precursor solution, ~100 µm thick spacers were used to prevent direct contact between liquid precursor and the photomask. The prepolymer layer was then exposed to UV light (60 mW/cm^2) through a chrome/soda lime photomask for ~0.5 s using UV source. UV exposure time leading to well-resolved hydrogel features requires optimization. Regions of PEG-DA exposed to UV underwent free-radical polymerization and became cross-linked, while unexposed regions were dissolved in DI water. A coverslip was carefully removed before development in DI water. This process resulted in formation of PEG hydrogel microwells with Ab-decorated bottom and nonfouling side walls. The immunofluorescent images are shown in Fig. 6.2A and B. The green fluorescence signals observed within the microwells demonstrate the localization of Abs at the underlying layer of the wells (Fig. 6.2A). The red fluorescence signals originated from cytokine binding on the bottom of the wells indicate the presence of Ab molecules in the cell attachment sites (Fig. 6.2B).

6.2.1.4 Fabrication of a microfluidic device

A microfluidic device was used in conjunction with microwell arrays to control flow conditions in order to isolate T-cells from a blood sample. Poly(dimethyl siloxane) (PDMS)-based microfluidic devices were fabricated using standard soft lithography approaches (Whitesides et al., 2001). Inlet/outlet holes were then punched with a blunt 16-gauge needle. The microfluidic device contained two flow chambers with width–length–height dimensions of 3 mm × 10 mm × 0.1 mm and a network of independently addressed auxiliary channels. The auxiliary channels were used to apply negative pressure (vacuum suction) to the PDMS mold and reversibly secure it on top of a glass substrate. This strategy allowed to seal a fluid conduit on top of the Ab-containing PEG hydrogel microwells without compromising immobilized biomolecules. A 5-ml syringe was connected to silicone tubing (1/32 in. I.D., Fisher), which was attached to the outlet of the flow chamber with a metal insert cut from a 20-gauge needle. A blunt, shortened 20-gauge needle carrying a plastic hub was inserted in the inlet. A pressure-driven flow in the microdevice was created by withdrawing the syringe positioned at the outlet with a precision syringe pump (Harvard Apparatus, Boston, MA).

6.2.1.5 Capturing T-cells and detecting secreted IFN-γ using sensing hydrogel microwells

In a T-cell capture experiment, RBC-depleted blood cells were seeded onto PEG gel microwells built on Ab spots. Prior to cell seeding, a PDMS device containing fluidic and vacuum channels was sterilized by 15 min of UV exposure in a tissue culture hood. The PDMS device was then aligned with the functionalized regions of the surface and sealed with negative pressure applied to the auxiliary channels. Afterward,

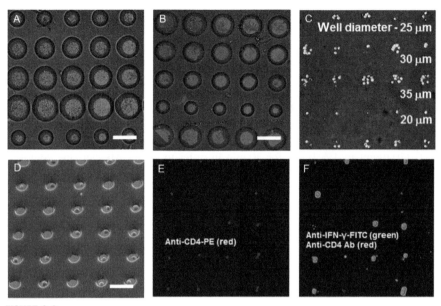

FIGURE 6.2

(A–D) PEG hydrogel microwells containing T-cell capture (anti-CD4) and cytokine detection (anti-IFN-γ) Abs. (A) Micropatterned surfaces were stained with FITC-labeled antimouse IgG to show localization of the Abs (green fluorescence) at the bottom of PEG hydrogel microwells. (B) To reveal presence of cytokine-sensing Abs, micropatterned surfaces were challenged with human recombinant IFN-γ (500 ng/ml) and then incubated with anti-IFN-γ-biotin and streptavidin-Alexa546 (red fluorescence). (C) Hydrogel microwells were enclosed in a microfluidic device and incubated with RBC depleted human blood, resulting in capture of the cells. (D) SEM image showing an array of single T-cells residing in 20-μm-diameter hydrogel microwells. (E, F). Detection of IFN-γ production from single T cells. Cells were captured on micropatterned surfaces inside a microfluidic device and then activated with mitogens for 15 h. (E) Staining with anti-CD4-PE (red) shows that all of the captured cells are CD4 positive T-cells. (F) The same image stained with anti-IFN-γ-biotin and neutravidin-FITC shows "halo" of green fluorescence due to secreted cytokine. Scale bar = 50 μm. (See color plate.)

sterile $1 \times$ PBS was injected into the flow chamber to remove air bubbles and then 50 μl of RBC-lysed blood resuspended in phenol red-free RPMI1640 media was added into the inlet reservoir (hub of a 20-gauge needle) and drawn into a microfluidic channel at a flow rate of 10 μl/min (0.3 dyn/cm^2). After the entry of cells into a microfluidic channel, the flow was stopped and cells were allowed to bind for 5–10 min. To wash away nonspecific cells, the flow rate was increased to between 50 (1.7 dyn/cm^2) and 100 μl/min (3.4 dyn/cm^2). The captured leukocytes were identified by immunofluorescent staining with FITC-labeled anti-CD3 Abs. Figure 6.2C clearly exhibits the capture of immune cells in Ab-coated PEG hydrogels. Importantly, as seen in Fig. 6.2C and D, 20-μm-diameter microwells resulted in single T-cell capture, whereas larger microwells captured multiple cells per well.

To detect IFN-γ secretion, T-cells captured in the microwells inside a microfluidic device were activated with a mitogenic solution comprised of PMA (50 ng/ml) and ionomycin (2 μM) in phenol red-free RPMI1640 media (with 10% FBS). After flushing mitogen with 1 × PBS, microfluidic chambers were filled with biotinylated anti-IFN-γ Abs (5 μg/ml in 1 × PBS) for 1 h at room temperature.

Next, the chambers were again flushed with 1 × PBS and filled with neutroavidin FITC (10 μg/ml in 1% BSA) for 30 min in order to detect Ab–cytokine complex (Fig. 6.2 F). Captured cells were stained with a PE-labeled anti-CD4 (1/10 dilution in 1% BSA) within the chambers. The fluorescently labeled cytokines and cells were then visualized and imaged with a confocal microscope. Figure 6.2E shows the immune cells stained positive for CD4 marker (red fluorescence), indicating CD4 T-cells being captured in PEG hydrogel microwells. Significantly, as shown in Fig. 6.2 F, immunofluorescent staining revealed a IFN-γ signal (green fluorescence) associated with individual T-cells.

6.2.2 Integration of sensing hydrogel microstructures into micropatterned co-cultures

This surface micropatterning strategy is a combination of two approaches: photoresist lithography-based micropatterning of surfaces (Bhatia et al., 1997) and hydrogel micropatterning. In both cases, glass substrates are first functionalized using a silane coupling agent described in Section 6.2.1.1. The moderately hydrophobic silanized surface facilitates physical adsorption of proteins, does not support attachment of epithelial cells such as hepatocytes but supports adhesion of fibroblasts and provides acrylate groups for covalent attachment of hydrogels. Sections below first describe protocols of photoresist lithography-based patterning of surfaces and then detail integration of resist and hydrogel lithography into the same process.

6.2.2.1 Photoresist lithography

Silane-modified glass slides were micropatterned by photoresist lithography to fabricate metal alignment marks (Fig. 6.3B). AZ5214-E positive photoresist was spin-coated on the silane-modified glass substrate at 800 rpm for 10 s followed by 4000 rpm for 30 s. The coated slide was soft-baked on a hot plate at 100 °C for 85 s. The photoresist layer was then exposed to UV light (10 mW/cm^2) for 35 s using a Canon PLA-501 F mask aligner. UV exposure time should be optimized based on the guidelines provided by the photoresist manufacturer. Exposed photoresist was then developed for 5 min in AZ 300 MIF developer solution, briefly washed with DI water to remove residual developing solution, and then dried under nitrogen. Development time needs to be optimized along with UV exposure time based on the guidelines provided by photoresist manufacturer. After washing, photoresist patterns could be observed under the

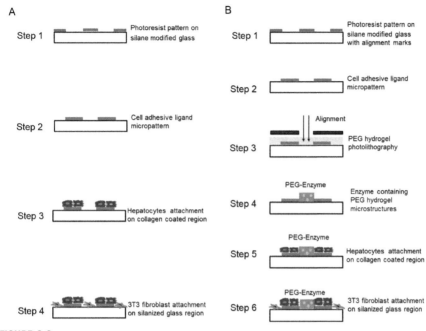

FIGURE 6.3

Creating micropatterned co-cultures. (A) Photoresist lithography to construct co-cultures. Step 1: Photoresist patterns were created on silane-modified surface by photoresist lithography. Step 2: Cell adhesive ligand (ECM protein) patterns are created on silane-modified surface after the removal of photoresist by lift-off. Step 3: Hepatocytes attach and form cell cluster on collagen (I) domains but not on silane-modified regions of the glass substrate. Step 4: Nonparenchymal cells (3 T3 fibroblasts) attach on silane-modified regions of the substrate, forming co-cultures. (B) Combining photoresist and hydrogel lithography to create co-cultures with integrated sensing elements. Step 1: Al alignment marks and photoresist patterns were created on silane-modified surface by photoresist lithography. Step 2: Cell adhesive ligand (ECM protein) patterns are created on silane-modified surface after the removal of photoresist by lift-off. Step 3: HRP-containing PEG prepolymer was exposed to UV through a photomask in alignment with Al marks. Step 4: Prepolymer not exposed to UV is removed by development in water. The surface contains hydrogel microstructures in registration with protein patterns. Step 5: Adult hepatocytes attach and form cell cluster on collagen (I) domains but not on silane-modified regions of the glass substrate. Step 6: Nonparenchymal cells (3 T3 fibroblasts) attach on silane-modified regions of the substrate, forming co-cultures. (For color version of this figure, the reader is referred to the online version of this chapter.)

microscope. A 500 Å aluminum layer was deposited onto glass slides with patterned photoresist using an E-beam evaporator (Fremont, CA). The glass slides were then placed into acetone and sonicated for 20 min to lift off the Al layer on top of the photoresist, thereby creating Al alignment marks.

The silanized glass slides containing Al alignment marks were micropatterned again with photoresist as described earlier. Registration of the photoresist layer with alignment marks was ensured with either a mask aligner or an upright microscope. It should be noted that alignment marks were required only in cases where hydrogel structures and protein patterning would need to be registered on the same surface. Alignment marks were not required when using photoresist lithography to prepare co-cultures without hydrogel structures.

6.2.2.2 Micropatterning proteins using photoresist layer as a stencil

To create collagen I micropatterns, a glass slide containing a photoresist pattern and alignment marks was incubated in a solution of collagen I (0.1 mg/ml in $1 \times$ PBS) for 30 min at room temperature. A slide was then briefly washed with DI water, sonicated in acetone for 10 min to remove the photoresist, and dried under nitrogen. The effect of acetone treatment on protein was investigated in terms of cell adhesion and cellular functions (Lee, Shah, Zimmer, Liu, & Revzin, 2008). In the case of collagen I, there was no detrimental effect of acetone sonication on cell adhesion. The glass slides containing collagen patterns can be kept at 4 °C for at least 1 month without detrimental effects on cell adhesion. For the visualization of the patterns, collagen I was labeled with Cy3 using a FluoroLink Cy3 reactive dye five pack based on the company's instructions.

6.2.2.3 Fabricating enzyme carrying PEG hydrogel microstructures in registration with collagen micropatterns

Photolithographic patterning of PEG hydrogel was described in Section 6.2.1.3. In the case of enzyme carrying PEG hydrogel microstructures, HRP stock solution was prepared by dissolving in $1 \times$ PBS to make the final concentration at 10 mg/ml. A prepolymer solution was prepared by mixing PEG-DA (MW 575) with 1% (w/v) photoinitiator (DMPA). Then HRP stock solution (10%, v/v) was mixed with PEG-DA prepolymer solution (90%, v/v). This HRP-containing precursor solution was spin-coated at 600 rpm for 5 s onto a glass surface. A photomask was placed on top of the liquid layer of PEG-DA precursor solution and aligned to alignment marks (indicators of collagen domains) using an upright microscope or a mask aligner. Then the sample was exposed through a chrome/soda lime photomask to UV for 1.5 s at 60 mJ/cm^2 using an OmniCure series 1000 light source or 10 s at 10 mJ/cm^2 using a Cannon PLA-501 F mask aligner.

6.2.2.4 Creating micropatterned co-culturing and detecting model metabolites

Prior to cell seeding, glass pieces were sterilized with 70% ethanol, washed twice with $1 \times$ PBS, and then placed into the wells of a conventional six-well plate. For cellular micropatterning, collagen/PEG hydrogel-patterned slides were exposed to 3 ml of primary rat hepatocyte suspension in culture medium at a concentration of 1×10^6 cells/ml. Hepatocytes require cell-adhesive ligands; therefore cell patterns formed on the collagen I coated glass regions (Fig. 6.3A, Step 3). Typically, hepatocytes start spreading out in 30 min. After 1 h of incubation at 37 °C, the medium containing unattached cells was removed and the surfaces were washed twice with 1 × PBS. To create co-cultures, glass slides containing hepatocytes were exposed to 3 ml of fibroblast cell suspension at a concentration of 5×10^5 cells/ml. After 15 min of incubation at 37 °C, the fibroblast culture medium was replaced with hepatocyte culture medium (DMEM supplemented with 10% FBS, 200 U/ml penicillin, 200 µg/ml streptomycin, 7.5 µg/ml hydrocortisone, 20 ng/ml epidermal growth factor, 14 ng/ml glucagon, and 0.5 U/ml insulin). Fibroblasts attached on moderately adhesive silanized glass regions (Fig. 6.3A, Step 4). Typical images of micropatterned surfaces are shown in Fig. 6.4. Photoresist pattern can be seen in Fig. 6.4A, this pattern serves as a stencil during adsorption of collagen (I) and can be removed by lift-off process using acetone. Figure 6.4B shows circular collagen-Cy3 domains created on the surface after adsorption of protein and removal of photoresist. Figure 6.4C and D shows surfaces after adsorption of hepatocytes (C) on collagen islands and subsequent attachment of fibroblasts on the surrounding glass regions (D). To better visualize co-cultures, hepatocytes and fibroblasts were labeled with Cell-Tracker red CMTPX (ex/em 577 nm/602 nm) and CellTracker green CMFDA (ex/em

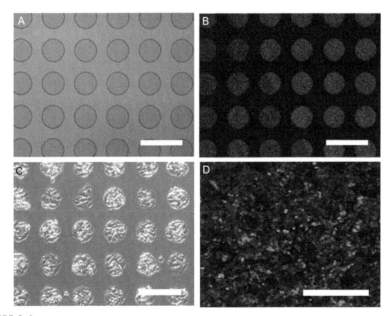

FIGURE 6.4

Micropatterned co-cultures created by photoresist lithography. (A) Photoresist pattern generated on silane-modified glass substrate. (B) Cy3-labeled collagen (I) pattern created by physisorption onto the open area of photoresist pattern. (C) Hepatocytes seeded on protein pattern generated well-defined cell clusters, which correspond to protein patterns in size and shape. (D) Two fluorescently labeled cells formed co-culture by successive seeding (red: hepatocytes, green: fibroblasts). Scale bar = 200 µm. (See color plate.)

492 nm/517 nm), respectively. As seen from Fig. 6.4D the two cell types were effectively segregated, hepatocytes in circular clusters surrounded by fibroblasts.

The same seeding protocol was followed when creating co-cultures with integrated sensing elements. These micropatterns combined three surface types: collagen (I) domains that are well suited for hepatocyte attachment, glass regions where hepatocytes do not attach but fibroblasts can attach, and hydrogel structures to which neither cell type can attach. Figure 6.5A shows one example of a micropatterned surface used for cell seeding. These surfaces contained collagen (I) domains (red fluorescence) inside hydrogel rings surrounded by glass. Hepatocytes incubated with such a surface preferred attachment on collagen pads inside the ring (Fig. 6.5B).

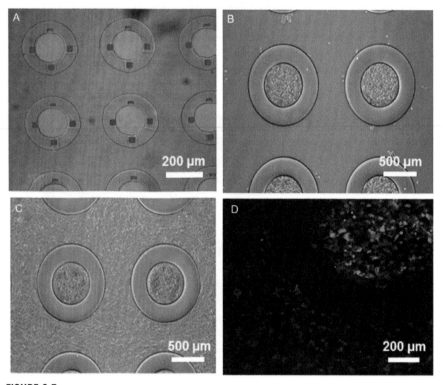

FIGURE 6.5

(A) Micropatterned surfaces containing collagen (I) regions (red fluorescence) in registration with hydrogel rings on silanized glass substrates. (B) After incubation with the micropatterned surface, rat hepatocytes selectively adhered to collagen domains, leaving glass regions open for attachment of "stickier" fibroblasts. (C) 3 T3 fibroblasts attached on silane-modified glass regions next to hepatocytes. PEG hydrogel structures did not support cell attachment during co-culture assembly. (D) Fluorescent labeling of cells showing a distinct population of hepatocytes (green) co-cultured with fibroblasts (red). PEG hydrogel features did not support cell attachment and are nonfluorescent. (See color plate.)

Once hepatocytes were given time to spread out on the surface and form cell–cell contacts, fibroblasts were added onto the same surface and attached on the outside of the ring (Fig. 6.5C). Once again, fluorescent labeling of cells was used to confirm that the two cell types were segregated on the surface (Fig. 6.5D). For this experiment hepatocytes and fibroblasts were labeled with CellTracker green CMFDA (ex/em 492 nm/517 nm) and CellTracker red CMTPX (ex/em 577 nm/602 nm), respectively (Fig. 6.5D).

For a proof of concept of experiment, Amplex Red reagent was used as a fluorescence probe to detect HRP-catalyzed oxidation of hydrogen peroxide (H_2O_2). This reagent is nonfluorescent in reduced form but becomes fluorescent when oxidized in the presence of H_2O_2 and HRP (Lee et al., 2009; Yan et al., 2009).

Amplex Red reagent was dissolved in DMSO to a final concentration of 5 mM and was stored in a desiccator at $-20\,^{\circ}C$. For detection of H_2O_2, Amplex red solution was added to the cell culture medium to create a concentration of 50 μM. Finally, to demonstrate enzymatic activity of HRP molecules entrapped in PEG hydrogel structures, H_2O_2 at a concentration of 10 μM was added to the cell culture medium. Figure 6.6 shows that HRP-containing PEG microstructures were sensitive to exogenous H_2O_2 in co-cultures. One should note that these experiments simply demonstrate that hydrogel structures can be integrated with co-cultures and be sensing the microenvironment. These experiments stop short however of detecting endogenous, cell-secreted factors being released or exchanged in co-cultures.

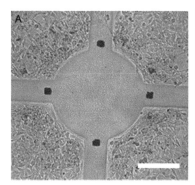 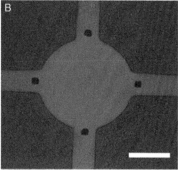

FIGURE 6.6

Integration of enzyme-containing PEG hydrogel structures with micropatterned co-cultures. Merged bright-field/fluorescence images of HRP-carrying hydrogel microstructures before (A) and after (B) addition of 10 μM H_2O_2. Strong red fluorescence observed in (B) points to retain activity of entrapped HRP. Scale bar = 100 μm. (For interpretation of the references to color in this figure legend, the reader is referred to the online version of this chapter.)

DISCUSSION AND CONCLUSION

Integration of sensing Abs with single cell culture system shown in Fig. 6.1 was developed to detect IFN-γ production of individual CD4 T-cells isolated from human blood. This strategy allowed us to create nonfouling PEG hydrogel microwells onto the cell- and cytokine-specific Abs-coated hydrogel layer. We investigated whether the coprinting of cell- and cytokine Abs has a detrimental impact on the sensitivity of IFN-γ (Zhu et al., 2009). The result revealed that the mixing IFN-γ Abs with CD4 Abs had no appreciable effect on the sensitivity of IFN-γ. Combining micropatterned sensing surfaces with microfluidics was particularly useful because of the small blood volumes required and precise control over washing conditions.

In another demonstration of the sensing micropatterns, enzyme-containing hydrogel microstructures were integrated with cells. Surface modification protocols were created to integrate sensing elements into precise locations within micropatterned co-cultures (Fig. 6.3B). In this method, PEG hydrogel microstructures served dual purpose of guiding cell attachment and sensing. HRP-containing PEG gel microstructures were found to be resistant to cell attachment, indicating that introduction of enzyme did not influence on nonfouling features of PEG structures (Lee et al., 2009). Additionally, we examined how long the encapsulated enzymes were retained inside the PEG structures using FITC-labeled dextran of similar molecular weight. The majority of enzyme molecules were retained in the PEG structures after 4 days in culture.

In subsequent studies we demonstrated the detecting H_2O_2 from macrophages or hepatocytes using optical or electrochemical hydrogel-based sensors (Matharu, Enomoto, & Revzin, 2013; Pedrosa, Enomoto, Simonian, & Revzin, 2011; Yan et al., 2009). We find electrochemical detection of secreted H_2O_2 described in references (Matharu et al., 2013; Pedrosa et al., 2011) to be more sensitive and reproducible compared to optical detection. It should be noted that while detection of secreted signals in monocultures has been shown and co-cultures with integrated biosensors have been constructed by us, the monitoring of signals being exchanged within the co-cultures has not yet been demonstrated. This represents one of the promising future directions that leverage the ability to place sensing elements into defined locations within cellular micropatterns.

References

Allcock, H., Phelps, M., Barrett, E., Pishko, M., & Koh, W. (2006). Photolithographic development of polyphosphazene hydrogels for potential use in microarray biosensors. *Chemistry of Materials*, *18*, 609–613.

Bhatia, S. N., Yarmush, M. L., & Toner, M. (1997). Controlling cell interactions by micropatterning in co-cultures: Hepatocytes and 3 T3 fibroblasts. *Journal of Biomedical Materials Research*, *34*, 189–199.

Borgmann, S., Radtke, I., Erichsen, T., Blçchl, A., Heumann, R., & Schuhmann, W. (2006). Electrochemical high-content screening of nitric oxide release from endothelial cells. *ChemBioChem, 7*, 662–668.

Cai, X., Klauke, N., Glidle, A., Cobbold, P., Smith, G. L., & Cooper, J. M. (2002). Ultra-low-volume, real-time measurements of lactate from the single heart cell using microsystems technology. *Analytical Chemistry, 74*, 908–914.

Co, C. C., Wang, Y. C., & Ho, C. C. (2005). Biocompatible micropatterning of two different cell types. *Journal of the American Chemical Society, 127*, 1598–1599.

Jung, S.-K., Gorski, W., Aspinwall, C. A., Kauri, L. M., & Kennedy, R. T. (1999). Oxygen microsensor and its application to single cells and mouse pancreatic islets. *Analytical Chemistry, 71*, 3642–3649.

Khademhosseini, A., Langer, R., Borenstein, J., & Vacanti, J. P. (2006). Microscale technologies for tissue engineering and biology. *Proceedings of the National Academy of Sciences of the United States of America, 103*, 2480–2487.

Lee, J. Y., Shah, S. S., Yan, J., Howland, M. C., Parikh, A. N., Pan, T., et al. (2009). Integrating sensing hydrogel microstructures into micropatterned hepatocellular cocultures. *Langmuir, 25*, 3880–3886.

Lee, J. Y., Shah, S. S., Zimmer, C. C., Liu, G.-Y., & Revzin, A. (2008). Use of photolithography to encode cell adhesive domains into protein microarrays. *Langmuir, 24*, 2232–2239.

Matharu, Z., Enomoto, J., & Revzin, A. (2013). Miniature enzyme-based electrodes for detection of hydrogen peroxide release from alcohol-injured hepatocytes. *Analytical Chemistry, 85*, 932–939.

Park, T. H., & Shuler, M. L. (2003). Integration of cell culture and microfabrication technology. *Biotechnology Progress, 19*, 243–253.

Pedrosa, V. A., Enomoto, J., Simonian, A. L., & Revzin, A. (2011). Electrochemical biosensors for on-chip detection of oxidative stress from immune cells. *Biomicrofluidics, 5*, 032008-1–032008-11.

Revzin, A., Rajagopalan, P., Tilles, A. W., Berthiaume, F., Yarmush, M. L., & Toner, M. (2004). Designing a hepatocellular microenvironment with protein microarraying and poly(ethylene glycol) photolithography. *Langmuir, 20*, 2999–3005.

Revzin, A., Russell, R. J., Yadavalli, V. K., Koh, W.-G., Deister, C., Hile, D. D., et al. (2001). Fabrication of poly(ethylene glycol) hydrogel microstructures using photolithography. *Langmuir, 17*, 5440–5447.

Revzin, A., Sekine, K., Sin, A., Tompkins, R. G., & Toner, M. (2005). Development of a microfabricated cytometry platform for characterization and sorting of individual leukocytes. *Lab on a Chip, 5*, 30–37.

Revzin, A., Tompkins, R. G., & Toner, M. (2003). Surface engineering with poly(ethylene glycol) photolithography to create high-density cell arrays on glass. *Langmuir, 19*, 9855–9862.

Russell, R. J., & Pishko, M. V. (1999). Poly(ethylene glycol) hydrogel-encapsulated fluorophore-enzyme conjugates for direct detection of organophosphorus neurotoxins. *Analytical Chemistry, 71*, 4909–4912.

Singhvi, R., Kumar, A., Lopez, G. P., Stephanopoulous, G. N., Wang, D. I. C., Whitesides, G. M., et al. (1994). Engineering cell-shape and function. *Science, 264*, 696–698.

Suh, K. Y., Seong, J., Khademhosseini, A., Laibinis, P. E., & Langer, R. (2004). A simple soft lithographic route to fabrication of poly(ethylene glycol) microstructures for protein and cell patterning. *Biomaterials, 25,* 557–563.

Takayama, S., McDonald, J. C., Ostuni, E., Liang, M. N., Kenis, P. J. A., Ismagilov, R. F., et al. (1999). Patterning cells and their environments using multiple laminar fluid flows in capillary networks. *Proceedings of the National Academy of Sciences of the United States of America, 96,* 5545–5548.

Troyer, K. P., Heien, M. L., Venton, B. J., & Wightman, R. M. (2002). Neurochemistry and electroanalytical probes. *Current Opinion in Biotechnology, 6,* 696–703.

Voldman, J., Gray, M. L., & Schmidt, M. A. (1999). Microfabrication in biology and medicine. *Annual Review of Biomedical Engineering, 1,* 401–425.

Whitesides, G. M., Ostuni, E., Takayama, S., Jiang, X., & Ingber, D. E. (2001). Soft lithography in biology and biochemistry. *Annual Review of Biomedical Engineering, 3,* 335–373.

Yan, J., Sun, Y., Zhu, H., Marcu, L., & Revzin, A. (2009). Enzyme-containing hydrogel micropatterns serving a dual purpose of cell sequestration and metabolite detection. *Biosensors and Bioelectronics, 24,* 2604–2610.

Zhu, H., Stybayeva, G., Silangcruz, J., Yan, J., George, M., Ramanculov, E., et al. (2009). Detecting cytokine release from single human T-cells. *Analytical Chemistry, 81,* 8150–8156.

Microfluidic Patterning of Protein Gradients on Biomimetic Hydrogel Substrates

Steffen Cosson, and Matthias P. Lutolf

Laboratory of Stem Cell Bioengineering, Institute of Bioengineering, School of Life Sciences, Ecole Polytechnique Fédérale de Lausanne, Lausanne, Switzerland

CHAPTER OUTLINE

Abstract

This protocol describes a versatile microfluidic method to generate tethered protein gradients of virtually any user-defined shape on biomimetic hydrogel substrates. It can be applied to test, in a microenvironment of physiologically relevant stiffness, how cells respond to graded biomolecular signals, for example to elucidate how morphogen proteins affect stem cell fate. The method is based on the use of microfluidic flow focusing to rapidly capture in a step-wise manner tagged biomolecules via affinity binding on the gel surface. The entire patterning process can be performed in <1 h. We illustrate one application of this method, namely, the spatial control of mouse embryonic stem cell self-renewal in response to gradients of the self-renewal-promoting signal leukemia inhibitory factor.

INTRODUCTION

Microfabrication techniques provide unprecedented means to probe and manipulate mammalian cells in culture (Kobel & Lutolf, 2010; Whitesides, 2006). Among the countless techniques that have emerged, micropatterning methods such as microcontact printing (Bernard, Renault, Michel, Bosshard, & Delamarche, 2000), inkjet printing (Boland, Xu, Damon, & Cui, 2006; Pardo, Wilson, & Boland, 2003), or microfluidics (Dertinger, Jiang, Li, Murthy, & Whitesides, 2002; Dertinger, Chiu, Jeon, & Whitesides, 2001; Jiang et al., 2005) are of particular interest because they enable to precisely control the spatial distribution and dose of desired cell-instructive signals. Microfluidic technology is particularly well suited to build *in vitro* model systems for the study of dynamic cellular processes such as cell migration, or multicellular morphogenetic processes, because this approach is ideally suited to fabricate *graded* biomolecule patterns (Keenan & Folch, 2008; Sant, Hancock, Donnelly, Iyer, & Khademhosseini, 2010).

However, state-of-the-art microfluidic gradient systems have some shortcomings for the analysis of difficult-to-culture cell types such as stem cells: First, long-range patterns with stable cross-sectional profile along the entire length of the pattern are difficult to obtain due to issues related to lateral diffusion of proteins within microscale channels, and lack of specificity in protein immobilization yields poor results when trying to overlap multiple patterns (Cosson, Kobel, & Lutolf, 2009). Second, cell culture at the microscale is still somewhat poorly characterized and cells within microchannels are often subjected to continuous flow that may induce shear stress-related issues (Toh & Voldman, 2011). Third, standard microfluidic devices are fabricated from poly(dimethylsiloxane) (PDMS) that despite its many advantageous properties (e.g., transparency, gas permeability, or biocompatibility) has shortcomings such as evaporation of medium, protein adsorption, or leaching of uncured monomer (Berthier, Young, & Beebe, 2012). Finally, substrate-bound protein gradients are often presented to cells on rigid glass or plastic substrate that do not provide an ideal substrate for stem cell culture.

To address these issues, we have developed microfluidic approaches to pattern protein gradients on the surface of soft and biomimetic poly(ethylene glycol) (PEG) hydrogels. Here, we describe one such method that is based on software-assisted hydrodynamic flow focusing for the controlled deposition of tagged proteins on the engineered hydrogels. We provide optimized patterning parameters to yield user-defined micropatterns with stable cross-sectional profile along the entire length of the pattern (>1 cm) and validate the method with a demonstration of a stem cell-based assay.

7.1 FABRICATION OF A HYDRODYNAMIC FLOW FOCUSING CHIP

Traditional microfabrication may not be accessible to most biological laboratories as it requires access to clean room facilities and some degree of microfabrication expertize. However, an increasing number of academic institutions have established foundry facilities that are open to external customers (e.g., Stanford University: http://www.stanford.edu/group/foundry/). For institutions that do not provide such microfabrication services commercial solutions exist (e.g., Micralyne or Micronit Microfluidics).

With respect to the protocol described below, we propose an inexpensive benchtop alternative to fabricate microfluidic chips by replica molding of master made out of cut Scotch tape (the reader is referred to the following publication: "Simple fabrication of microfluidic devices by replicating Scotch-tape master," Chips & Tips, *Lab on a Chip*, 17 May 2010). Indeed, as the microfluidic layout is rather simple with smallest features below 300 μm, the chip can be readily fabricated using this simple soft-lithography fabrication process.

7.2 GENERATION OF THIN HYDROGEL FILMS ON SILANIZED GLASS SLIDES

Here, we present a detailed protocol for the fabrication of thin PEG hydrogel films necessary for the rapid capture of proteins from solution.

7.2.1 Materials

- Dynasylan® MTMO 3-trimethoxysilypropane-1-thiol (MPS) (Degussa GmbH Aerosil&Silanes)
- Ethanol absolute, ACS grade (Fisher, United Kingdom)
- Acetic acid, ACS grade (Ridel de Haën, Germany)
- Hydrogel precursor macromers: 8 arm-PEG-vinyl sulfone (10 kDa, PEG-VS, synthesized as described elsewhere (Lutolf & Hubbell, 2003)) and 4 arm-PEG-thiol (10 kDa, PEG-TH, NOF Corporation, Japan).

- PEGylated protein A. The PEGylation of recombinant protein A (Biovision, USA) is described elsewhere (Cosson et al., 2009).
- NeutrAvidin-maleimide (Thermo Scientific, USA)
- Dithiothreitol (DTT), ACS grade (BioRad, USA)
- Triethanolamine buffer (0.3 M, pH 8, Fluka, USA).

7.2.2 Equipment

- Oxygen plasma cleaner femto (Diener electronic GmbH + Co, Germany).

7.2.3 Silanization of glass slides

Glass slides (silica surface) were modified by a treatment with 3-mercaptopropyl-trimethoxylsilane (MPS) to generate free thiol groups to generate a stable interface between the hydrogel and the glass surface (Fig. 7.1B, scheme 1). The silanization was adapted from Huang, Nair, Nair, Zingaro, and Meyers (1995).

7.2.4 Method

1. Clean the glass slides with detergent, bi-distilled water, and ethanol.
2. Dry on air.
3. Oxygen plasma the glass slides for 5 min.
4. Mix 300 ml of ethanol, 3 ml of MPS, and 10 drops ($\approx$0.3 ml) of acetic acid.

Note: In the presented protocol we preferred ethanol to toluene to solubilize the MPS for toxicity reasons.

5. Immerse the glass slides in the solution for 30 min.
6. Rinse with 70% ethanol and dry on air.
7. Bake the treated glass slides in an oven at 110 °C for 1 h.
8. Store the slides at 4 °C.

Note: Immerse the MPS-treated glass slides for 5–10 min in a 10 mM DTT solution prior to use in order to break disulfide bonds that may have formed.

7.2.5 Hydrogel casting

Here, we describe the fabrication of a thin PEG hydrogel layer on silanized glass slides. Of note, the thiol-modified gelatin Gelin-S (Glycosan) that was chosen as adhesion ligand for mouse embryonic stem cell (ESC) (Figs. 7.2A and 7.3) used in step 5 may be replaced by other biomolecules containing a free thiol.

1. Mix hydrogel precursors to yield 20 μl of 5% w/v hydrogel with equimolar ratio of functional groups ([VS]/[TH] = 1), containing 3.36 mg/ml of PEGylated protein A and/or NeutrAvidin-maleimide.

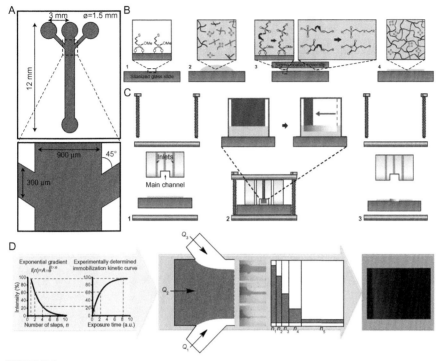

FIGURE 7.1

Hydrogel patterning by hydrodynamic flow focusing. (A) Hydrodynamic flow focusing microfluidic layout. (B) Engineered hydrogel casting. 1. Prepare a silanized glass slide. 2. Pipette a 20 μl drop of the hydrogel premix on the silanized glass slide. 3. Cover the drop with a Sigma-coated coverslip to spread the drop evenly. Put in an incubator at 37 °C for 15 min to allow gelation. 4. Delicately remove the Sigma-coated coverslip and recover the thin hydrogel layer on the silanized glass slide. (C) Microfluidic device assembly. 1. Press the PDMS chip on the glass slide bearing the thin hydrogel layer and tighten the chip holder. 2. Connect the device to an automated syringe pump and pattern the hydrogel by hydrodynamic flow focusing. 3. Release the patterned hydrogel from the microfluidic device. (D) Hydrogel patterning by hydrodynamic flow focusing. 1. Program the automatic syringe pump by correlating your mathematical model to the experimentally determined immobilization kinetic curve. 2. Pattern the hydrogel in a succession of discrete steps of defined duration and spatial area covered by the protein stream. 3. Recover the patterned hydrogel for imaging or cell-based assay. (See color plate.)

Note: The total amount of binding agent may be raised up to a total of 6.64 mg/ml, however higher concentrations may lead to poor hydrogel gelation and delamination from the glass substrate.

2. Pipette the 20 μl on the surface of a Sigma-coated coverslip (2 cm × 3 cm) and place it upside down onto a MPS-treated glass slide.

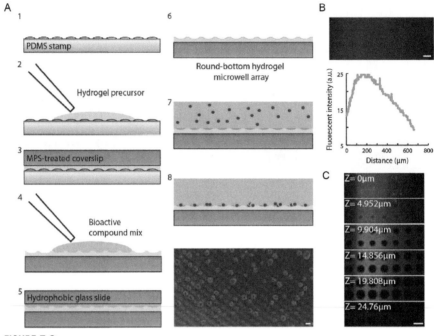

FIGURE 7.2

Micropatterning of topographically structured hydrogel. (A) Schematic representation of a topographically structured hydrogel. 1. PDMS template for round-bottom micro-well molding. 2. Pipette a drop of hydrogel premix on the PDMS template. 3. Place a MPS-treated coverslip on the hydrogel premix solution to spread it and place it in an incubator at 37 °C for 15 min. 4. Gently remove the PDMS template to recover the topographically structured hydrogel layer. Pipette a drop of bioactive compound, that is, Gelin-S, on the hydrogel. 5. Place a Sigma-coated coverslip on the drop to spread it and place it in an incubator at 37 °C for 1 h. 6. Remove the coverslip and wash thoroughly the hydrogel with PBS. Store the topographically structured biofunctional hydrogel immersed in PBS at 4 °C, at least overnight prior to use. 7. Place the hydrogel in a multiwell plate, add a mES cell suspension (i.e., 15,000 cell per well), and place it in an incubator at 37 °C for cell culture. 8. mES will adhere to the hydrogel surface and form colonies preferentially within the round-bottom micro-wells. (B) Fluorescent micrograph and graphical representation of a linear gradient of fluorescently labeled bovine serum albumin on a topographically structured hydrogel. (C) Confocal fluorescent micrographs from a Z scan (scale bar = 100 µm). (See color plate.)

3. Place it in an incubator at 37 °C for 15 min.
4. Carefully remove the Sigma-coated coverslip.
5. Pipette 200 µl of 0.2% Gelin-S solution (http://www.glycosan.com/products/gelin-s/) over the hydrogel surface and place it in the incubator for 1 h.

Note: Spread the drop with a pipette tip such that the whole hydrogel surface is recovered.

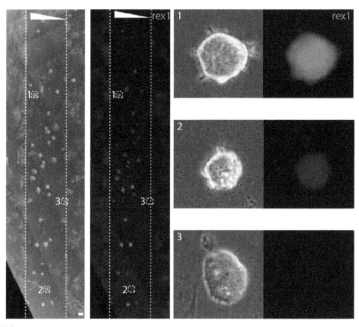

FIGURE 7.3

Directed self-renewal of mouse embryonic stem cells on tethered FcLIF gradients. Bright field and fluorescent micrographs of embryonic stem cells culture for 3 days on tethered linear FcLIF gradients. White dashed lines delimit the gradient area. Enlarged micrographs of regions of interest (numbered dashed white boxes) (scale bar = 100 μm). (For color version of this figure, the reader is referred to the online version of this chapter.)

6. Wash the hydrogel three times with phosphate-buffered saline (PBS).

7. Store the hydrogels immersed in PBS at 4 °C for more than 4 h to enable hydration of the hydrogel.

7.3 BIOMOLECULE MICROPATTERNING ON HYDROGELS

Here, we present a detailed protocol for patterning biomolecules on hydrogel substrates prepared above.

7.3.1 Materials

- PDMS hydrodynamic flow focusing chip
- Chip holder consisting of two metal plates (4 cm × 5 cm) with centered holes (1.5 cm × 1.5 cm) for visualization, and four screws and nuts (m3)
- Engineered hydrogel on silanized glass slide prepared as described earlier
- PBS (1 ×, Gibco, USA)
- Syringe 1 ml (Omnifix 1 ml, BBraun, USA)

- Tygon tubing
- Blunt needle (precision tips 23 G, Nordson EFD, USA)

7.3.2 Equipment

- Desiccator and vacuum pump
- Oven
- Software driven syringe pump (e.g., NEMEsys, Cetoni GmbH, Germany)
- Fluorescent microscope

7.3.3 Method

7.3.3.1 Assembly of microfluidic setup

1. Place the hydrogel-coated glass slide on the chip holder (Fig. 7.1C, scheme 1).
2. Press the PDMS chip on the hydrogel and tighten the chip holder (Fig. 7.1C, scheme 2).
3. Prime the microchannels with PBS using a Pasteur pipette to remove any air bubbles.

Note: We may immerse the device in PBS under vacuum for 30 min if air bubbles remain trapped in the microchannels.

4. Fill two syringes (1 ml) with PBS. Fill a syringe with a solution of a desired Fc-tagged protein (0.1 mg/ml).
5. Connect the syringes with Tygon tubing with the help of a blunt needle.
6. Mount the syringes on the automated syringe pump.
7. Connect the tubing to the PDMS chip. The two PBS-filled syringes on the side channel inlets and the syringe containing Fc-tagged protein on the central inlet.
8. Connect a Tygon tubing to the outlet to remove excess waste liquid.

7.3.3.2 Hydrogel patterning

1. Program the automatic syringe pump. Parameters for gradient patterning of various profiles (linear, exponential, or Gaussian) can be obtained for each discrete patterning step by (i) correlating the desired profile to the fitted empiric immobilization curves (Fig. 7.1D, equations of the fitted immobilization curves for both immobilization scheme are provided in Table 7.1) and (ii) calculate (see Table 7.2) the proper input flow rates from the equations below to control the protein stream width.
2. Run the program and monitor the patterning process under a microscope.

Note: If the protein to be tethered is not fluorescently labeled, it may be necessary to add fluorescent proteins in the first syringe filled with PBS to facilitate visualization of the patterning process.

Table 7.1 Fitted immobilization kinetic curves[a]

	Fitted curve equation	a	b	c	d	R^2
NeutrAvidin/ BSA-biotin	$f(x) = a \times e^{bx} + c \times e^{dx}$	3986	0.001683	−3976	−1.377	0.9976
Protein A/IgG	$F(x) = a \times e^{bx} + c \times e^{dx}$	−3453	−0.1592	3584	0.0048	0.9954

Note: In this graphs the intensities span from 0 to 4095.
[a]The presented equations were obtained by fitting the immobilization curves empirically obtained for our model proteins at a given hydrogel functionalization (3.36 mg/ml of capturing agent) and at a given flow rate of the protein stream (15 µl/min) which can be found elsewhere (Allazetta, Cosson, & Lutolf, 2011).

Table 7.2 Flow rate calculations

Step	Q_1 (µl/min)	Q_2 (µl/min)	Q_3 (µl/min)
1	3	15	12
n	3	$Q_{2,n-1} \times (d/n_{max} \times w_n)$	$25 - Q_{1,n-1} - Q_{2,n-1}$

d is the main channel width (ca. 900 µm), n_{max} is the number of discrete patterning step, and w_n is the desired protein stream width at each discrete patterning step. Note that we keep the total flow rate constant (ca. 25 µl/min), which was found to yield minimal lateral diffusion from the central protein stream.

3. Once the patterning program is finished set the pump to 10 µl/min for both syringe filled with buffer and perfuse the chip for 5 min to remove unbound proteins.
4. Unplug the chip.
5. Under a flow hood, dismantle the device and recover the patterned hydrogel.
6. Wash it three times with PBS and place it in a 12-well plate. Add 1 ml of PBS and store at 4 °C before imaging or cell-based experiments.

7.4 DISCUSSION

Microfluidic technology is a powerful means to generate continuous, high-resolution patterns of tethered biomolecules. This protocol described how to generate such micropatterns on soft and biomimetic hydrogels. That is to say, the use of a hydrogel substrate may be advantageous because it allows performing cell culture experiments in the context of microenvironments whose physicochemical properties are reminiscent of natural cellular microenvironments. Furthermore, the hydrogel characteristics (e.g., stiffness, presentation of adhesion ligands, or signaling cues) can be readily tailored for cell-specific requirements. Finally, the removal of the microfluidic setup from the substrate after patterning affords the culture of cells in traditional multiwell plate formats.

7.4.1 Pitfalls and drawbacks

The equipment necessary to use our method (e.g., chip fabrication equipment, syringe pumps) is rather costly. We have therefore proposed an alternative method for the fabrication of microfluidic chip (see Section 7.1) which does not require access to a clean room facility. Moreover, instead of a programmable syringe pump that we employ for hydrodynamic flow focusing, the use of hydrostatic pressure driven flow may be considered. Finally, particular caution is required to avoid any possible contamination during the patterning process, which has to be performed outside of a sterile environment.

7.4.2 Perspectives and outlook

In the following section, we highlight possible extensions of the technique and we also provide a proof-of-principle example of its application to influence the fate of ESCs.

7.4.2.1 Generation of overlapping biomolecule gradient using orthogonal binding schemes

In the described protocol we provide two orthogonal protein immobilization schemes, namely biotin/NeutrAvidin and Fc-tag/protein A. Rather large amounts of proteins (max. 200 ng/cm^2) with similar resolution can be micropatterned either singly or in combination (Cosson, Allazetta, & Lutolf, 2013). The introduction of this orthogonal binding scheme enables reliable overlapping of two proteins in various configurations (parallel antisense, parallel superimposed, and even orthogonally crossing).

7.4.2.2 Micropatterning of topographically structured hydrogels

Microwell arrays embossed in hydrogel are advantageous for stem cell biology (e.g., Lutolf, Doyonnas, Havenstrite, Koleckar, & Blau, 2009). Microfluidic micropatterning method may not be optimal to pattern microwell arrays due to limitations related to the microwell geometry. To overcome this issue, we use round-bottom hydrogel microwell arrays which allow compartmentalizing/arraying cell colonies as shown for ESCs in Fig. 7.2A.

Notably, using the same micropatterning protocol uniform patterning of round-bottom microwell surfaces (Fig. 7.2B and C) can be achieved, thanks to the laminar flow conditions within microchannels. Although the depth of the microwells used for this example is small (ca. 13–15 μm), we believe that deeper wells can be patterned and this method should be useful for various cell-based assays.

7.4.2.3 Example: controlling ESC fate via graded self-renewal cues (Fig. 7.3)

ESC self-renewal can be spatially controlled by presenting a gradient of hydrogel-tethered Fc-tagged leukemia inhibitory factor (FcLIF) (Cosson et al., 2013). The micrographs in Fig. 7.3 show that self-renewing colonies, marked here by the green fluorescent protein (GFP) that reports the expression of the pluripotency marker Rex1/Zfp42, are clustered on the area of the hydrogel where high concentrations of FcLIF are presented. On areas with low tethered FcLIF, as well as nonpatterned areas, differentiation of ESC is observed as is evident from altered morphologies and loss of GFP expression.

CONCLUSIONS

This protocol highlights a versatile microfluidic-based method for generating biomolecule patterns on biomimetic hydrogel substrates. The hydrogel substrate can be tailor-made to have physical and biochemical characteristics that are specific to a particular cell type and cell function of interest. The use of orthogonal biomolecule binding schemes affords generating signaling microenvironments that are rather complex, for example, overlapping protein gradients. Furthermore, the use of topographically structured gel surfaces can be employed to arrange cells into distinct colonies which can be useful for a number of applications.

References

Allazetta, S., Cosson, S., & Lutolf, M. P. (2011). Programmable microfluidic patterning of protein gradients on hydrogels. *Chemical Communications*, *47*(1), 191–193. http://dx. doi.org/10.1039/C0cc02377a.

Bernard, A., Renault, J. P., Michel, B., Bosshard, H. R., & Delamarche, E. (2000). Microcontact printing of proteins. *Advanced Materials*, *12*(14), 1067–1070. http://dx.doi.org/10.1002/1521-4095(200007)12:14<1067::Aid-Adma1067>3.0.Co;2-M.

Berthier, E., Young, E. W. K., & Beebe, D. (2012). Engineers are from PDMS-land, biologists are from polystyrenia. *Lab on a Chip*, *12*(7), 1224–1237. http://dx.doi.org/10.1039/C2lc20982a.

Boland, T., Xu, T., Damon, B., & Cui, X. (2006). Application of inkjet printing to tissue engineering. *Biotechnology Journal*, *1*(9), 910–917. http://dx.doi.org/10.1002/biot.200600081.

Cosson, S., Allazetta, S., & Lutolf, M. P. (2013). Patterning of cell-instructive hydrogels by hydrodynamic flow focusing. *Lab on a Chip*, *13*(11), 2099–2105. http://dx.doi.org/10.1039/c3lc50219h.

Cosson, S., Kobel, S. A., & Lutolf, M. P. (2009). Capturing complex protein gradients on biomimetic hydrogels for cell-based assays. *Advanced Functional Materials*, *19*(21), 3411–3419.

Dertinger, S. K., Jiang, X., Li, Z., Murthy, V. N., & Whitesides, G. M. (2002). Gradients of substrate-bound laminin orient axonal specification of neurons. *Proceedings of the National Academy of Sciences of the United States of America*, *99*(20), 12542–12547. http://dx.doi.org/10.1073/pnas.192457199.

Dertinger, S. K. W., Chiu, D. T., Jeon, N. L., & Whitesides, G. M. (2001). Generation of gradients having complex shapes using microfluidic networks. *Analytical Chemistry*, *73*(6), 1240–1246.

Huang, L., Nair, P. K., Nair, M. T. S., Zingaro, R. A., & Meyers, E. A. (1995). Chemical deposition of bi2s3 thin films on glass substrates pretreated with organosilanes. *Thin Solid Films*, *268*(1–2), 49–56.

Jiang, X., Xu, Q., Dertinger, S. K., Stroock, A. D., Fu, T. M., & Whitesides, G. M. (2005). A general method for patterning gradients of biomolecules on surfaces using microfluidic networks. *Analytical Chemistry*, *77*(8), 2338–2347. http://dx.doi.org/10.1021/ac048440m.

Keenan, T. M., & Folch, A. (2008). Biomolecular gradients in cell culture systems. *Lab on a Chip*, *8*(1), 34–57. http://dx.doi.org/10.1039/b711887b.

Kobel, S., & Lutolf, M. (2010). High-throughput methods to define complex stem cell niches. *Biotechniques*, *48*(4), ix–xxii. http://dx.doi.org/10.2144/000113401.

Lutolf, M. P., Doyonnas, R., Havenstrite, K., Koleckar, K., & Blau, H. M. (2009). Perturbation of single hematopoietic stem cell fates in artificial niches. *Integrative Biology (Cambridge)*, *1*(1), 59–69. http://dx.doi.org/10.1039/b815718a.

Lutolf, M. P., & Hubbell, J. A. (2003). Synthesis and physicochemical characterization of end-linked poly(ethylene glycol)-co-peptide hydrogels formed by Michael-type addition. *Biomacromolecules*, *4*(3), 713–722. http://dx.doi.org/10.1021/bm025744e.

Pardo, L., Wilson, W. C., & Boland, T. J. (2003). Characterization of patterned self-assembled monolayers and protein arrays generated by the ink-jet method. *Langmuir*, *19*(5), 1462–1466. http://dx.doi.org/10.1021/La026171u.

Sant, S., Hancock, M. J., Donnelly, J. P., Iyer, D., & Khademhosseini, A. (2010). Biomimetic gradient hydrogels for tissue engineering. *Canadian Journal of Chemical Engineering*, *88*(6), 899–911.

Toh, Y. C., & Voldman, J. (2011). Fluid shear stress primes mouse embryonic stem cells for differentiation in a self-renewing environment via heparan sulfate proteoglycans transduction. *FASEB J*, *25*(4), 1208–1217. http://dx.doi.org/10.1096/fj.10-168971.

Whitesides, G. M. (2006). The origins and the future of microfluidics. *Nature*, *442*(7101), 368–373. http://dx.doi.org/10.1038/nature05058.

Cell and Tissue Micropatterning in 3D

Micropatterning of Poly(ethylene glycol) Diacrylate Hydrogels

Saniya Ali, Maude L. Cuchiara, and Jennifer L. West

Department of Biomedical Engineering, Duke University, Durham, North Carolina, USA

CHAPTER OUTLINE

Abstract

This protocol describes the techniques to synthesize and fabricate micropatterned poly(ethylene glycol) diacrylate-based hydrogels that can be used as substrates in cellular studies and tissue engineering scaffolds. These materials provide an essentially bioinert background material due to the very low protein adsorption characteristics of poly(ethylene glycol), but the materials can be modified with covalently grafted

Methods in Cell Biology, Volume 121   ISSN 0091-679X
http://dx.doi.org/10.1016/B978-0-12-800281-0.00008-7

peptides, proteins, or other biomolecules of interest to impart specific biofunctionality to the material. Further, it is possible to use micropatterning technologies to control the localization of such covalent grafting of biomolecules to the hydrogel materials, thus spatially controlling the cell–material interactions. This protocol presents a relatively simple approach for mask-based photolithographic patterning, generally best suited for patterning the surface of hydrogel materials for 2D cell studies. A more sophisticated technique, two-photon laser scanning lithography, is also presented. This technique allows free-form, 3D micropatterning in hydrogels.

INTRODUCTION

With the emergence of substrate rigidity as an important factor influencing many aspects of cellular behavior, there has been increasing interest in cell culture substrates with mechanical properties that can be tuned within a physiologically relevant range. Poly(ethylene glycol) diacrylate (PEGDA)-based hydrogel materials are a biocompatible substrate that offer this mechanical tunability (Nemir, Hayenga, & West, 2010) as well as the ability to micropattern in both 2D (Hahn, Taite, et al., 2006b; Moon, Hahn, Kim, Nsiah, & West, 2009; Leslie-Barbick, Shen, Chen, & West, 2011) and 3D (Hahn, Miller, & West, 2006a; Hoffmann & West, 2010; Lee, Moon, & West, 2008).

Poly(ethylene glycol) (PEG) is a hydrophilic, neutral, and highly mobile polymer chain that has shown very good biocompatibility in a wide range of biomedical applications (Ratner, 2004). PEG is highly resistant to protein adsorption (Harris, 1992), and thus hydrogels formed by cross-linking of PEG are generally cell nonadhesive. The bioinert nature of PEG hydrogels acts as a blank slate upon which one can engineer desired biological functionality (Fig. 8.1). For example, covalent attachment of cell adhesive peptides renders PEG hydrogels cell adhesive (Hern & Hubbell, 1998), but if photolithographic patterning is employed to spatially limit the covalent attachment of cell adhesive peptides, cell binding to the hydrogel surfaces can be confined to desired regions (Hahn, Taite, et al., 2006b).

PEGDA is one type PEG derivative that is easily fabricated into hydrogels via addition polymerization of the acrylate termini (Sawhney, Pathak, & Hubbell, 1993). Biocompatible photoinitiators along with long wavelength UV or visible light (depending on choice of initiator) can be used to induce cross-linking. The use of photochemistry is beneficial because PEGDA hydrogels can be formed very quickly, making cell encapsulation feasible; it provides temporal control over cross-linking, easing handling; and many spatial patterning technologies are enabled by photochemistry. Below we provide information for mask-based photolithographic patterning of PEGDA hydrogels, best suited for patterning surfaces of hydrogels used to seed cells for 2D studies, as well as a newer patterning technique, two photon laser scanning lithography, which allows micropatterning in 3D and can be used in hydrogels with cells seeded throughout for 3D culture studies. This can allow investigators to study things such as 3D cell motility in response to immobilized gradients of

FIGURE 8.1

A PEGDA hydrogel. This hydrogel was formed by photocross-linking an aqueous solution of PEGDA. PEGDA hydrogels are transparent, highly hydrated, and can be easily fabricated into a variety of sizes and shapes. (For color version of this figure, the reader is referred to the online version of this chapter.)

various bioactive factors, cell fate determination in response to 3D patterns of morphogens, and cross-talk between biochemical signaling and mechanotransduction, as well as to potentially form increasingly complex engineered tissues.

8.1 SYNTHESIS OF PEGDA

8.1.1 Materials

PEG of desired molecular weight (Fig. 8.2)
Acryloyl chloride
Triethylamine (TEA)
Ultrahigh purity argon gas
Anhydrous dichloromethane (DCM)
DCM
1.5 M K_2CO_3
Anhydrous $MgSO_4$
Diethyl ether.

8.1.2 Equipment

Three-neck round-bottom flask
Glass stopcock
Separatory funnel
Beakers (various sizes)

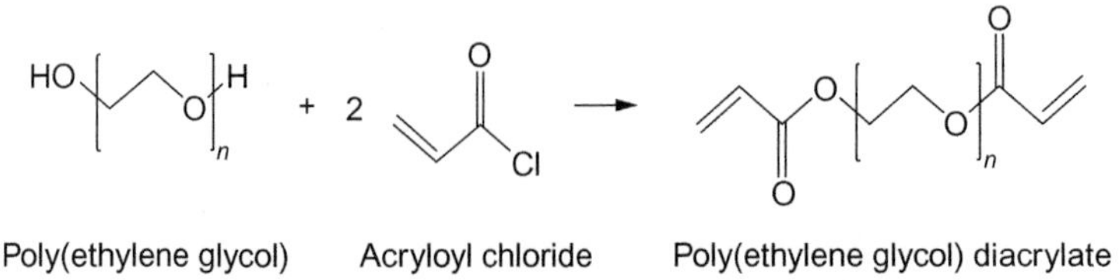

Poly(ethylene glycol)   Acryloyl chloride   Poly(ethylene glycol) diacrylate

FIGURE 8.2

Synthesis of PEGDA. Polyethylene glycol of the desired molecular weight is reacted with acryloyl chloride.

Glass syringes
Deflected point septum penetration needles
Glass funnels
Erlenmeyer vacuum flasks
Filter paper (#5, Whatman, Piscataway, NJ)
Büchner funnels
Round-bottom flask
Crystallization dish
Aluminum foil.

8.1.3 Method

1. Lyophilize PEG of desired molecular weight and mass overnight.
2. Clean all glassware in a base bath and dry overnight in the 100 °C oven.
3. In a fume hood, set up a three-neck round-bottom flask with one stopcock connected to vacuum and argon gas (Ar) and two septa.
4. Evacuate and inert flask with Ar three times.
5. Add PEG with Ar still flowing into the flask. Evacuate and inert with Ar three times.
6. Use a glass syringe to add 10 ml increments of anhydrous DCM to the PEG. Stir the mixture. Add DCM until all PEG is dissolved, and the solution is not viscous (~40 ml DCM per 24 g PEG).
7. Evacuate and inert with Ar three times.
8. Add TEA (2 mol TEA:1 mol PEG) to PEG solution with glass syringe. Evacuate and inert with Ar three times.
9. Allow solution to mix for 5 min.
10. Add acryloyl chloride to PEG solution with glass syringe *very slowly* (4 mol acryloyl chloride:1 mol PEG). Evacuate and inert with Ar three times.
11. Protect reaction from light. React at least 12 h.
12. In a fume hood, transfer the reacted mixture in the round-bottom flask to a separatory funnel.
13. Rinse round-bottom flask with two 10 ml volumes DCM to recover PEG.
14. Add 1.5 M K_2CO_3 to the separatory funnel. Use a volume that is 1/4–1/2 of the total volume in the separatory funnel. Reducing this volume will prevent PEGDA from going into the aqueous phase.

15. Stopper and shake the funnel vigorously for a few seconds, venting the funnel to release CO_2 between shakings. Repeat until no more gas is released. Solution should have the consistency of a milky white emulsion.
16. Remove the glass stopper and replace it with parafilm.
17. Protect from light. Allow the phases to separate by gravity overnight.

Note: The mixture in the separation funnel should have separated into two phases. The organic phase containing most of the PEGDA is denser and thus the bottom phase. If left long enough ($\sim$20 h), three phases can be seen: an upper aqueous phase (clear), a middle organic phase (white), and a lower organic phase (off white) containing the PEGDA.

18. In a fume hood, drain the lowest organic phase into a beaker being careful not to let any other phase drain out. Stir the mixture.
19. While stirring, add anhydrous $MgSO_4$ until the mixture goes from a lumpy consistency to a well-dispersed mixture of powder and organic solvent. (If too much $MgSO_4$ is added, DCM can be added to drive the PEGDA into solution.)
20. Prewet filter paper with DCM. Vacuum filter mixture through a Büchner funnel to remove $MgSO_4$. Rinse beaker with DCM and filter to recover the maximum amount of PEGDA.
21. Transfer the filtered solution to a round-bottom flask. Rinse the vacuum flask with additional DCM and add to round-bottom flask. (Round-bottom flask should be less than half full.)
22. Rotary evaporate the solution to concentrate the PEGDA. Remove enough DCM so that the solution just begins to become viscous.
23. Precipitate PEGDA out of solution in diethyl ether. While stirring the ether, pour in PEGDA solution. PEGDA will precipitate as a white crystal. Let the mixture stir for 10 min to move residual TEA to the organic phase.
24. Prewet filter with ether. Filter through Büchner funnel to collect PEGDA.
25. Dry thoroughly in a crystallization dish overnight in the fume hood and then under a vacuum. *Note*: Make sure the PEGDA does not smell strongly of ether before placing in a vacuum. Dry PEGDA may be crushed with a mortar and pestle to get a finer powder.
26. Store powder at $-20\,^{\circ}C$ until ready to use. Expect a $\sim$75% recovery.

Note: The purified PEGDA should be characterized by ^{1}H-NMR. PEGDA has a characteristic peak at $\sim$4.32 ppm corresponding to the PEG protons adjacent to the acrylate group.

8.2 SYNTHESIS OF ACRYLOYL-PEG-CONJUGATED PEPTIDES

As an example, the cell adhesion peptide Arginine-Glycine-Aspartic Acid-Serine (RGDS) is used in this synthesis protocol. This can be easily replaced with another peptide sequence as desired with little impact on the protocol. The reagent used reacts with amine groups, so peptide sequences with additional primary amines that are necessary for their bioactivity can be negatively impacted. Alternative reactive PEG reagents, such as maleimide derivatives, are commercially available and should be explored in such situations. Additionally, this protocol is designed to modify a peptide with a single PEG-acrylate. It is also possible to design a peptide with a lysine residue (or multiple lysines) to create additional sites for PEG-acrylate attachment. This can be done to create proteolytically degradable PEG-diacrylate derivatives that enable cell migration through the hydrogel materials (Moon et al., 2010; West & Hubbell, 1998). In such cases, the same protocol can be followed, but the molar ratios in the reaction steps should be adjusted relative to the number of primary amines in the peptide sequence.

8.2.1 **Materials**

Acrylate-PEG-succinimidyl valerate (PEG-SVA), molecular weight $= 3400$ g/mol (Laysan Bio, Arab, Alabama)
RGDS peptide (American Peptide, Sunnyvale, CA)
Anhydrous dimethyl sulfoxide (DMSO)
N,N-Diisopropylethylamine (DIPEA)
Ultra high purity argon.

8.2.2 **Equipment**

Four milliliters amber vial
Glass syringes
Deflected point septum penetration needles
Regenerated cellulose dialysis tubing (Spectrum Labs, MWCO 3500)
Forty milliliters amber vial
Vortexer
Lyophilizer.

8.2.3 **Method**

1. Bring PEG-SVA and a 20-mg vial of RGDS to room temperature prior to opening.
2. Add 0.5–1.0 ml DMSO into RGDS vial. Vortex the vial until the RGDS has completely dissolved.
3. Transfer the solution to a 4-ml amber vial.
4. Rinse the RGDS vial with two additional 0.5 ml volumes of DMSO, vortexing and transferring each to the 4-ml amber vial for a total volume of 1.5–2.0 ml.
5. Add PEG-SVA (1 mol PEG-SVA: 1.2 mol RGDS) directly to the amber vial with the RGDS in DMSO.
6. Add DIPEA to the vial (1 mol DIPEA: 2 mol RGDS).

7. Immediately cap the vial and vortex on high for 3 min. Fill with argon, parafilm, and vortex until all PEG-SVA is fully dissolved. A total of 1–2 ml DMSO can be added to help solubilize the PEG-SVA if necessary.

8. Wrap the reaction vial in foil and place on rocker plate at room temperature for a minimum of 4 h.

9. Add 10-ml ultrapure water to a 40-ml amber vial and chill on ice for 10 min.

10. Add the contents of the reaction vial drop-wise to the ice cold water.

11. Rinse the reaction vial with 0.5-ml ultrapure water and transfer to the 40-ml vial. Cap and mix gently.

12. Transfer the contents to a dialysis membrane that has been rinsed. Rinse the vial two more times with 2-ml ultrapure water and add that to the membrane as well.

13. Dialyze against 4-l ultrapure water with at least four changes of water, at least 1–2 h apart. Keep the container completely covered with aluminum foil during dialysis.

14. Freeze the product at −80 °C and lyophilize. Store under argon at −20 °C.

Note: The product should be analyzed by gel permeation chromatography or mass spectrometry to check conjugation efficiency. Expect conjugation efficiencies of 85–100%. Analysis by nuclear magnetic resonance is also recommended. If desired, a fluorescent acryloyl-PEG-RGDS can be easily prepared by reacting this product with AlexaFluor 488 carboxylic acid tetrafluorophenyl (TFP) ester (see Section 8.3).

8.3 MASK-BASED PHOTOLITHOGRAPHIC PATTERNING

PEG-RGDS is used as an example for patterning in this protocol. This can be easily substituted for any other mono-acrylated PEG derivative. In the case of PEG-RGDS, this patterning process would yield regions that were cell adhesive (due to the presence of the grafted RGDS) against a cell nonadhesive background, as shown in Fig. 8.3.

8.3.1 Materials

PEGDA
PEG-RGDS
HEPES buffered saline (HBS) pH 7.4
Phosphate buffered saline (PBS) pH 7.4
Sodium bicarbonate buffer (50 mM) pH 8.5
AlexaFluor 488 carboxylic acid, TFP ester (Life Technologies, Grand Island, NY)
2,2-Dimethoxy-2-phenyl-acetophenone (DMAP)
N-Vinylpyrrolidinone (NVP).

8.3.2 Equipment

Adobe Illustrator
Transparency sheets
Regenerated cellulose dialysis tubing (Spectrum Labs, MWCO 3500)
Glass slides

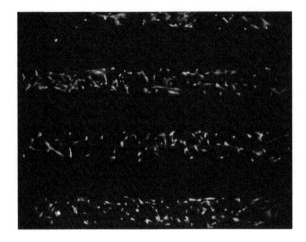

FIGURE 8.3

Cells grown on the surface of a hydrogel patterned with the RGDS peptide. The mask-based photolithographic patterning process described in Section 8.3 was used to form stripes of immobilized RGDS on the surface of a PEGDA hydrogel. Since the base hydrogel material was cell nonadhesive, cell adhesion was confined to the areas where RGDS was grafted to the hydrogel, as defined by the photomask that was used. (For color version of this figure, the reader is referred to the online version of this chapter.)

Teflon spacers (desired thickness)
Binder clips
Plastic syringes
Poly(ether sulfone) 0.22-μm syringe filters (EMD Millipore, Darmstadt, Germany)
Needles
Pipettor
Pipette tips
Long wavelength UV lamp (B-200SP UV lamp, UVP, 365 nm, 10 mW/cm^2)
Vortexer
Lyophilizer.

8.3.3 Method

1. Design desired patterns in Adobe Illustrator. This process is quite flexible with regards to pattern geometry.
2. Print at 20,000 dpi onto transparency masks. This can also be done through a vendor like CAD/ART Services.

   Note: If desired, chromium masks can be used in place of transparencies.

3. React Alexa Fluor 488 carboxylic acid, TFP ester with PEG-RGDS (10 mol dye:1 mol PEG-RGDS) for 1 h at room temperature in sodium bicarbonate buffer to create a fluorescently tagged PEG-RGDS molecule. Dialyze and lyophilize the final product.

4. Assemble hydrogel molds by sandwiching a Teflon spacer of the desired thickness between two glass slides and clamping with binder clips.

Note: If materials are intended for use in cell culture, sterilize at least 4 h under UV irradiation in a laminar flow hood and then perform steps 5–15 aseptically.

5. Dissolve PEGDA in 10-mM HBS to final concentration of 0.1 g/ml. Syringe filter the solution to sterilize. Add 10 μl/ml of 300 mg/ml DMAP in NVP and vortex.
6. Add the polymer solution to the molds using a syringe with needle or pipette. Use sufficient volume to create hydrogel of desired dimensions.
7. Cross-link under the long wavelength UV lamp for 30 s.
8. Soak PEGDA hydrogels in sterile PBS to swell overnight.

Note: After soaking and if desired, hydrogels can be cut into various sizes. Cork borers are good for creating hydrogel disks of uniform sizes.

9. Prepare 30 μmol/ml PEG-RGDS solution in HBS. Spike the solution with the fluorescently tagged PEG-RGDS (1–5% w/v can be adjusted experimentally to obtain desired fluorescent intensity).
10. Add 10 μl/ml of 300 mg/ml DMAP in NVP to PEG-RGDS solution and vortex. Pass the solution through a syringe filter to sterilize.
11. Place a PEGDA hydrogel onto a sterile glass slide. Pipette a thin layer of PEG-RGDS solution (estimate ~5 μl/cm^2) onto the PEGDA hydrogel surface.
12. Lay the patterned transparency mask on top of the precursor solution with the printed side facing the hydrogel surface. Place a Teflon spacer around the hydrogel and add a glass slide on top. Clamp the slides together.
13. Expose the hydrogel to long wavelength UV light through the transparency mask for 1 min.
14. Rinse the hydrogel with sterile PBS to remove unbound PEG-RGDS.
15. Transfer the hydrogel to sterile PBS and soak overnight.

Note: Patterns can be visualized on a fluorescent microscope. For cell studies, soak hydrogels in media prior to seeding.

8.4 TWO-PHOTON LASER SCANNING LITHOGRAPHY FOR 3D MICROPATTERNING

This process allows free-form 3D patterning within preformed PEGDA hydrogels with resolution down to 1 μm (Hoffmann & West, 2010). This protocol uses PEG-RGDS as an example. This can be easily substituted for other mono-acrylated PEG derivatives. If the molecular weight of the derivative we select is significantly different from PEG-RGDS, we may need to alter the incubation time in step 2 to optimize diffusion into the hydrogel. PEGDA is used as the base hydrogel in this

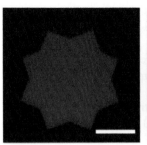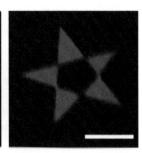

FIGURE 8.4

Three-dimensional micropatterns of RGDS within PEGDA hydrogels. The process described in Section 8.4 was utilized to form various shapes of RGDS within PEGDA-based hydrogels. Images of the patterns were acquired via confocal microscopy. Note that the base hydrogel is nonfluorescent and thus appears black. Patterns can easily be made in any user-defined shape desired. (For color version of this figure, the reader is referred to the online version of this chapter.)

example. This can be easily substituted for proteolytically degradable PEG-diacrylate derivatives without altering the protocol (Fig. 8.4).

8.4.1 Materials

MilliQ Water
Sterile 60 mm × 15 mm plastic Petri dishes
PBS pH 7.4
Glass slides
2,2-Dimethoxy-2-phenylacetophenone (DMAP)
NVP
PEGDA hydrogel (see Section 8.3, steps 4–8)
PEG-RGDS
PEG-RGDS-Alexa Fluor488 (see Section 8.3, step 3).

8.4.2 Equipment

Olympus FluoView1000 Multiphoton system equipped with a sample holder, Prior H117 ProScan III encoded XY motorized stage (Prior Scientific Inc., London, UK), FV10-ASW Olympus Fluoview software, tunable femtosecond Ti:Saph laser (680–1080 nm) with pulse compensation, long working distance objectives optimized for multiphoton (25 ×/1.05 numerical aperture water objective) (Olympus Corporation, Tokyo, Japan).

Custom-built chamber for incubation of hydrogels consisting of a Sigmacoat-treated (Sigma-Aldrich, St. Louis, MO) glass slide with a 500 μm layer of polydimethylsiloxane (Dow Corning, Midland, MI), with a 0.75 cm × 0.75 cm square cut in the center.

Optical Powermeter and Si Wand Detector, 400–1100 nm range (Newport, Irvine, CA)Platform rocker.

8.4.3 **Method**

1. Fill the custom-built incubation chamber with 50–100 µl of PEG-RGDS at the desired concentration, spiked with PEG-RGDS-AlexaFluor488 if desired for visualization of patterns, in PBS 10 µl/ml of 300 mg/ml DMAP in NVP.

Note: If materials are intended for cell culture, then reagents should be sterile, materials should be sterilized by exposure to UV, and processes should be carried out aseptically.

2. Place a PEGDA hydrogel into the incubation chamber, cover with glass and soak for 2 h at room temperature.

Note: If desired, cells can be suspended in the PEGDA solution prior to photocross-linking for encapsulation in the hydrogel material. Cells are generally used at densities ranging from 1 to 30 million cells/ml, depending on cell type and the purpose of our study. Depending on the purpose of our study, it may be advisable to utilize a proteolytically degradable derivative of PEGDA. If there are encapsulated cells in the hydrogel, keep in a cell culture incubator (37 °C, 5% CO_2) during the soak step.

3. Power up the Olympus multiphoton microscope system.

Note: Contact an experienced confocal user or Olympus representative for assistance and training if needed.

4. In the FluoView software, open the MP Laser Controller-Chameleon window and tune the laser to 720 nm.
5. Activate PMTs in the image acquisition control window by selecting a detector-RXD1 to 4. Make sure that AutoHV is selected and the PMTs are in Analog int mode. Set the adjustor gain (called HV in the Olympus Software-high voltage on the PMT), and the offset. The gain is just a digital multiplier so it is set to 1.
6. Acquire a single 2D image by pressing the XY acquire button. A blank image should appear in the 2D views window.
7. To create a configuration for patterning, select the rectangle, ellipse, line, or polygon regions of interest (ROI) tab to manually draw on the 2D image shown in the Live View window. Use as many ROIs as necessary in order to create the desired configuration. The ROI manager window allows for changing the x and y position, height and width, rotation angle, and laser power setting for each ROI.

8. Once the ROIs have been defined and a patterning configuration has been created, define the pattern settings (% laser power, scan speed/pixel dwell time, number of iterations, scan mode) in the acquisition setting window.

9. Run a series of power output measurements (in mW) of the laser using the optical power meter for various pattern settings. The software allows for defining % laser power in the laser panel of the acquisition setting which correlates to a certain watt of power output of the laser.

Note: This step is important to consider in order for determining the appropriate laser energy necessary to cross-link the desired amount of the ligand into the hydrogel. The amount of the ligand patterned into the material is highly dependent on the laser properties specifically laser power and scan speed/pixel dwell time. Generally, higher laser power and slower scan speed or longer dwell time yields greater concentration of the ligand cross-linked into the hydrogel.

10. Once an appropriate pattern setting is defined, position the sample on Prior H117 ProScan III encoded XY motorized stage and add a droplet of water on top of the sample. Adjust the stage so that the objective contacts that water on the sample.

11. Click on the Trans Lamp button in the software and turn on the Olympus LG-PS2 bulb to use transmitted light for visualization. Pull out both sliders on the left of the scope to use the eyepieces to locate and focus into the sample.

12. Once the sample has been focused, locate a region within the sample to pattern and define a z-stack by focusing up and down into the sample and setting up the start and end of the z-stack in the acquisition setting of the software. Select step size and number of slices for the z-stack.

13. Turn off the LG-PS2 bulb unit and push in both sliders on the left of the scope before starting to scan. Turn off the lights and place a photonic isolation device around the microscope.

Note: The photonic isolation device can be as simple as cardboard as long as it encases the optical path.

14. For scanning in the defined z-stack, click "Depth" below the XY acquire button to enable z-stack acquisition and click the acquisition button to start the scan. At the end of the scan, click "Series Done" to end the scan.

15. Remove the sample from the stage and place the patterned sample in a petri dish immersed with 10–20 ml of PBS. The patterned sample is washed extensively with PBS under gentle rocking for 48 h to remove unbound PEG-RGDS and photointiator with PBS under gentle rocking for 48 h, changing the PBS periodically.

Note: The resultant 3D micropatterns can be visualized by conventional confocal microscopy. This process can be done iteratively to pattern multiple ligands.

8.5 DISCUSSION

The majority of studies of cell responses to substrate rigidity have been performed on polyacrylamide hydrogels for 2D studies or else in collagen, fibrin, or Matrigel for 3D studies. Polyacrylamide has appropriately tunable mechanical properties and can be micropatterned (Wong, Velasco, Rajagopalan, & Pham, 2003), but the monomer is highly toxic, limiting its utility and application to 2D *in vitro* studies. Protein-based hydrogels, such as collagen, fibrin, and Matrigel, generally allow excellent cell adhesion and migration. However, when attempting to alter their mechanical properties, either by changing protein concentration or by cross-linking, the bioactivity is altered, confounding the interpretation of the results. Additionally, micropatterning is generally not possible, or at least very difficult, with these materials.

We believe a distinct advantage of the PEGDA-based materials is the ability to directly translate between 2D and 3D experiments. Numerous studies in the literature have noted distinct differences in cellular behaviors in 2D versus 3D (Baker & Chen, 2012; Cukierman, Pankov, Stevens, & Yamada, 2001; Fraley et al., 2010). It has often been difficult to interpret these observed differences, though, as in most instances, the 2D and 3D studies are carried out with completely different materials, with different mechanics and biological properties. Since cells can be either seeded on the surface of preformed PEGDA hydrogels for 2D studies or suspended in PEGDA solutions prior to photocross-linking for 3D studies, it is quite straightforward to use the same materials and cells for both 2D and 3D studies.

The use of two-photon excitation processes to micropattern these hydrogels in 3D dramatically advances the capabilities of the experimental systems one can develop. Based upon the same principle as two-photon microscopy, simultaneous absorption of two photons from a high-frequency pulsed laser to move a molecule (in this case, the photoinitiator rather than a fluorophore) into an excited state, illumination conditions can be set to limit excitation to a very small focal volume, leaving all other points in the laser path unaffected (Denk, Strickler, & Webb, 1990). This has previously been applied to PEGDA hydrogels to form complex shapes and gradients of immobilized factors (Hahn, Miller, et al., 2006a), to immobilize multiple peptides in adjacent patterns (Hoffmann & West, 2010), and in image-guided patterning (Culver et al., 2012).

GENERAL CONCLUSIONS

The patterning methods described in this protocol can be adapted for use with many different materials that undergo addition polymerization or are photoreactive. Other reactive PEG derivatives that have been widely used include PEG dimethacrylate and PEG fumarate. Photoreactive hydrogels have also been formed from other materials including hyaluronic acid, polyvinyl alcohol, alginate, chitosan, and dextran. An alternate strategy for two-photon micropatterning has been to modify hydrogels with coumarin-caged thiols, which can be locally uncaged under two-photon irradiation conditions to generate reactive thiol groups in 3D micropatterns, and these are subsequently reacted with maleimide-derivatized biomolecules (Wylie et al., 2011). There is considerable flexibility in the design parameters for these hydrogel cell culture systems and many new innovations are emerging.

References

Baker, B. M., & Chen, C. S. (2012). Deconstructing the third dimension—How 3D culture environments alter cellular cues. *Journal of Cell Science*, *125*, 1–10.

Cukierman, E., Pankov, R., Stevens, D. R., & Yamada, K. M. (2001). Taking cell-matrix adhesions to the third dimension. *Science*, *294*(5547), 1708–1712.

Culver, J. C., Hoffmann, J. C., Poche, R. A., Slater, J. A., West, J. L., & Dickinson, M. E. (2012). Three-dimensional biomimetic patterning in hydrogels to guide cellular organization. *Advanced Materials*, *24*(17), 2344–2348.

Denk, W., Strickler, J. H., & Webb, W. W. (1990). Two-photon laser scanning fluorescence microscopy. *Science*, *248*(4951), 73–76.

Fraley, S. I., Feng, Y., Krishnamurthy, R., Kim, D. H., Celedon, A., Longmore, G. D., et al. (2010). A distinctive role for focal adhesion proteins in three-dimensional cell motility. *Nature Cell Biology*, *12*(6), 598–604.

Hahn, M. S., Miller, J. S., & West, J. L. (2006a). Three-dimensional biochemical and biomechanical patterning of hydrogels for guiding cell behavior. *Advanced Materials*, *18*(29), 2679–2684.

Hahn, M. S., Taite, L. J., Moon, J. J., Rowland, M. L., Ruffino, K. A., & West, J. L. (2006b). Photolithographic patterning of polyethylene glycol hydrogels. *Biomaterials*, *27*(12), 2519–2524.

Harris, J. M. (1992). *Poly(ethylene glycol) chemistry: Biotechnical and biomedical applications*. New York: Plenum Press (385 p.).

Hern, D. L., & Hubbell, J. A. (1998). Incorporation of adhesion peptides into nonadhesive hydrogels useful for tissue resurfacing. *Journal of Biomedical Materials Research*, *39*(2), 266–276.

Hoffmann, J. C., & West, J. L. (2010). Three-dimensional photolithographic patterning of multiple bioactive ligands in poly(ethylene glycol) hydrogels. *Soft Matter*, *6*(20), 5056–5063.

Lee, S. H., Moon, J. J., & West, J. L. (2008). Three-dimensional micropatterning of bioactive hydrogels via two-photon laser scanning photolithography for guided 3D cell migration. *Biomaterials*, *29*(20), 2962–2968.

Leslie-Barbick, J. E., Shen, C., Chen, C., & West, J. L. (2011). Micron-scale spatially patterned, covalently immobilized vascular endothelial growth factor on hydrogels accelerates endothelial tubulogenesis and increases cellular angiogenic responses. *Tissue Engineering Part A*, *17*(1–2), 221–229.

Moon, J. J., Hahn, M. S., Kim, I., Nsiah, B. A., & West, J. L. (2009). Micropatterning of poly(ethylene glycol) diacrylate hydrogels with biomolecules to regulate and guide endothelial morphogenesis. *Tissue Engineering Part A*, *15*(3), 579–585.

Moon, J. J., Saik, J. E., Poche, R. A., Leslie-Barbick, J. E., Lee, S. H., Smith, A. A., et al. (2010). Biomimetic hydrogels with pro-angiogenic properties. *Biomaterials*, *31*(14), 3840–3847.

Nemir, S., Hayenga, H. N., & West, J. L. (2010). PEGDA hydrogels with patterned elasticity: Novel tools for the study of cell response to substrate rigidity. *Biotechnology and Bioengineering*, *105*(3), 636–644.

Ratner, B. D. (2004). *Biomaterials science: An introduction to materials in medicine*. Amsterdam: Elsevier Academic Press (851 p.).

Sawhney, A. S., Pathak, C. P., & Hubbell, J. A. (1993). Bioerodible hydrogels based on photopolymerized poly(ethylene glycol)-co-poly(α-hydroxy acid) diacrylate macromers. *Macromolecules*, *26*(4), 581–587.

West, J. L., & Hubbell, J. A. (1998). Polymeric biomaterials with degradation sites for proteases involved in cell migration. *Macromolecules*, *32*(1), 241–244.

Wong, J. Y., Velasco, A., Rajagopalan, P., & Pham, Q. (2003). Directed movement of vascular smooth muscle cells on gradient-compliant hydrogels. *Langmuir*, *19*(5), 1908–1913.

Wylie, R. G., Ahsan, S., Aizawa, Y., Maxwell, K. L., Morshead, C. M., & Shoichet, M. S. (2011). Spatially controlled simultaneous patterning of multiple growth factors in three-dimensional hydrogels. *Nature Materials*, *10*(10), 799–806.

Curved and Folded Micropatterns in 3D Cell Culture and Tissue Engineering

Cem Onat Yilmaz*, Zinnia S. Xu[†], and David H. Gracias[‡,§]

**Department of Materials Science and Engineering, Johns Hopkins University,
Baltimore, Maryland, USA*
[†]Department of Biomedical Engineering, Johns Hopkins University, Baltimore, Maryland, USA
*[‡]Department of Chemical and Biomolecular Engineering, Johns Hopkins University,
Baltimore, Maryland, USA*
[§]Department of Chemistry, Johns Hopkins University, Baltimore, Maryland, USA

CHAPTER OUTLINE

Abstract

Cells live in a highly curved and folded micropatterned environment within the human body. Hence, there is a need to develop engineering paradigms to replicate these microenvironments in order to investigate the behavior of cells *in vitro*, as well as to develop bioartificial organs for tissue engineering and regenerative medicine. In this chapter, we first motivate the need for such micropatterns based on anatomical considerations and then survey methods that can be utilized to generate curved and folded micropatterns of relevance to 3D cell culture and tissue engineering. The methods surveyed can broadly be divided into two classes: top-down approaches inspired by conventional 2D microfabrication and bottom-up approaches most notably in the self-assembly of thin patterned films. These methods provide proof of concept that the high resolution, precise and reproducible patterning of cell and matrix microenvironments in anatomically relevant curved and folded geometries is possible. A specific protocol is presented to create curved and folded hydrogel micropatterns.

INTRODUCTION

In the human body, there are on the order of 100 trillion (10^{14}) cells and thus spatially organizing cells while providing ample perfusion of nutrients and removal of waste is a major challenge (Hall, 2010). One could imagine packing all these cells in a clump but that would severely limit intercellular diffusion and transport of nutrients. Rather, in the human body, collections of cells are clustered in organs that are highly patterned in all three dimensions. Remarkably, no cell is more than 25–50 µm away from a capillary through which fluids flow and aid transport of nutrients to and waste away from cells. In addition to fluids, about 25 trillion red blood cells facilitate transport of oxygen through tubular blood vessels in the arteriovenous system which range in size from approximately 8 µm sized capillaries to the 2.5 cm sized aorta or the 3 cm sized vena cava (Berger, Goldsmith, & Lewis, 1996). Again in the context of transport, many organs such as the kidney and the lung have high surface areas for enhanced liquid or gas exchange. Consequently, an important anatomical feature in the human body is the presence of curved and micropatterned tissues such as capillaries, ducts, glomeruli, and alveoli, which are composed of layers of cells (e.g., endothelial, epithelial, fibroblasts) and extracellular matrix (e.g., collagen).

In addition to enabling efficient transport, curved and folded micropatterns also enhance interconnectivity between cells. For example, in the human brain, there are about a 100 billion (10^{11}) neurons, the largest number of cortical neurons as compared to all mammals. Notably, neurons in the cortex have on average approximately 30,000 synapses (Herculano-Houzel, 2012; Rockland, 2002). Further, the information processing capacity necessitates connections to be short requiring unique 3D architectures to facilitate this high level of interconnectivity. Curved and folded anatomical features such as gyri and sulci are prominent anatomical features of the brain.

The entire human body self-assembles these complex morphologies from the repeated cellular division of a single cell, the fertilized ovum. So this single cell must build the entire body through growth, cell division, and hierarchical morphogenesis, during which cells multiply as well as deposit and sculpt their extracellular microenvironments. Recent experiments in tissue morphogenesis (Eiraku et al., 2011; Sato & Clevers, 2013) show dramatic curving and folding of tissue such as the exvagination of an epithelial vesicle in an embryonic stem cell culture system

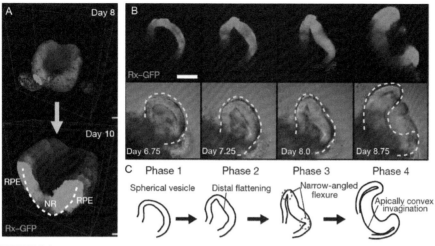

FIGURE 9.1

Progressive morphogenetic changes of ES-cell-derived retinal epithelium. (A) Surface-rendering 3D reconstruction images of invagination. The front part (bright green) is a cross-section image. (B) Optical cross-section of Rx–GFP images; Scale bar: 100 μm (top) and laser-scanned bright-field images (bottom; dotted lines indicate the basal side). (C) Four phases during the invagination process. (See color plate.)

Reprinted with permission from Eiraku et al. (2011), © 2011, Macmillan Publishers Ltd.

(Fig. 9.1). The multitude of recent findings of such self-folding and self-curving experiments are in agreement with prior understanding on growth, remodeling, and morphogenesis (Taber, 1995). Further these findings exemplify how a cell can multiply to form a spheroid and then generate exvaginations or invaginations to fold it into a 3D structure and in doing so give rise to the kinds of patterns which facilitate enhanced transport and interconnectivity in compact spaces.

In order to create *in vitro* models that accurately recapitulate the behavior of cells in their native environment within the human body as well as to create replacement organs and tissues, it is necessary to develop methods to pattern cells, biomaterials, and biomolecules in the widely observed curved and folded 3D geometries. In this chapter, we first detail a few examples of these anatomical geometries and architectures to highlight relevant features, sizes, and scales. We then review top-down and bottom-up methods that can be used to pattern curved and folded structures for cell culture and tissue engineering. We end with a summary and our future perspectives.

9.1 **ANATOMICAL CONSIDERATIONS**

The human anatomy has evolved to satisfy certain functions that are essential for its survival. The structure of components, whether organelles, cells, tissues, or organs in the body, is very closely correlated with function on many size scales, from proteins (Hegyi & Gerstein, 1999) to organs such as the kidney (Moriya, Tanaka, & Moriya, 2000). Even minor alterations in these biological structures have been linked to disease states. For example, protein damage that changes its molecular structure or misfolding can cause diseases such as amyloidosis, a disorder of the kidney caused by abnormal aggregation of proteins (Dobson, 2003; Hartl, 1996; Shirahama & Cohen, 1967). Also, changes that cause glomerular hypertrophy, an increase in the size of glomeruli, which are structures in the kidney involved with filtration, lead to hyperfiltration in diabetes mellitus type I, which can progress to chronic kidney disease (Helal, Fick-Brosnahan, Reed-Gitomer, & Schrier, 2012; Moriya et al., 2000; Rasch, Lauszus, Thomsen, & Flyvbjerg, 2005; Sasson & Cherney, 2012).

Space in the human body is limited and in order to fit trillions of cells in an organized fashion while maintaining all the necessary functions for survival, nutrient absorption, and waste removal (Shingleton, 2010), the need for patterns of curved and folded tissues in the body is essential. Since many of the most important functions in the human body occur at the interface between two materials, by opting for a repeated folded pattern, the surface area can be increased significantly to allow for many more functions per unit area. As examples, we focus on two structures, the glomeruli in the kidney, which are an example of a curved anatomical structure and the valvulae conniventes, villi and microvilli in the small intestine, which provide an example of a folded anatomical structure. While they have different physiological functions in the body, they have similar requirements that influence their structure.

9.1.1 Curved micropatterns in the kidney

While the kidney is a small organ, weighing less than 1% of an average adult human (Abrahams, 2007; Nyengaard & Bendtsen, 1992), its functions are indispensable to human survival—not only does it remove toxins and wastes from the blood, but more importantly it maintains homeostatic regulation of water and ions (i.e., Na^+, H^+) in the blood (Silverthorn, 2004). The amount of plasma processed by the nephrons each day is an incredible 180 l, compared to the total plasma volume (3 l) or the average volume of urine leaving the body per day (1.5 l/day) (Silverthorn, 2004). Since the kidney is limited by its small size, to achieve this amount of blood filtration per day, there needs to be significant intertwining and curving of the capillaries in order to greatly increase the surface area for volume exchange. Blood that needs to be filtered enters the kidney and goes through the glomeruli, coiled networks of fenestrated capillaries with large pores, allowing the contents of plasma to flow out to be filtered (Bergstrand & Bucht, 1958). Each glomerulus is surrounded by a round Bowman's capsule, forming a renal corpuscle, and serves as a tiny filtration unit (Abrahams, 2007), that removes plasma-like fluid from the capillaries into the renal tubules (Silverthorn, 2004). Many corpuscles are patterned in the renal cortex so that all the blood entering the kidneys is filtered. The glomerular size is positively correlated with body surface and adjusts the increased demand by increasing glomerular size and therefore surface area for exchange (Nyengaard & Bendtsen, 1992).

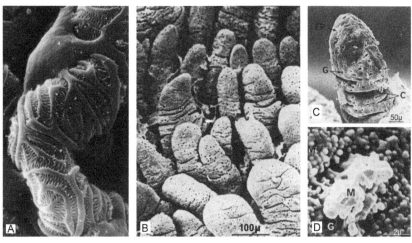

FIGURE 9.2

(A) Podocyte (rat) morphology via scanning electron microscopy (7300×). The podocytes wrap around the glomeruli capillaries with major processes. Foot-like projections extend off the processes and interlace with neighboring projections to create slit diaphragms where filtration occurs. Reprinted with permission from Mundel and Kriz (1995), © 1995, Springer. (B) Intestinal villi under electron scanning microscope (50×). Reprinted with permission from Marsh and Swift (1969), © 1996, BMJ Publishing Group Ltd. (C) One villus under electron scanning microscope (200×). The surface corrugations (C), epithelial cells (EP), and goblet cell orifices (G) can be seen. Reprinted with permission from Marsh and Swift (1969), © 1969, BMJ Publishing Group Ltd. (D) Electron scanning microscopic image of microvilli (MV). A goblet cell (G) secreted mucus (M) can also be seen.

Reprinted with permission from Marsh and Swift (1969), © 1969, BMJ Publishing Group Ltd.

Taking a closer look at the glomeruli capillaries, it can be seen that they are covered in a layer of podocytes (Fig. 9.2A) which are cells with foot-like projections that wrap and interlace around the capillaries (Mundel & Kriz, 1995). The spaces between the podocyte feet, known as slit diaphragms or filtration slits, are highly regulated as this is where blood filtration takes place (Arakawa, 1971; Höhne et al., 2012; Mundel & Kriz, 1995; Silverthorn, 2004; Wickelgren, 1999). These podocytes are responsible for the size restrictive filtration of molecules, as well as regulation of glomeruli filtration rate—a measure of renal function (Wickelgren, 1999), which is partially dictated by the surface area and permeability of the glomerular capillaries. Since the kidney serves such an important function of homeostatic regulation of water and ions in the blood, small changes in the sizes or shapes of the structures in the kidney can have a drastic effect. For example, narrowed afferent arterioles in the kidney can manifest as hypertension (Nørrelund et al., 1994).

9.1.2 **Folded micropatterns in the small intestines**

The small intestine is responsible for the absorption of nutrients from our diet (Kararli, 1995). Its main method of absorption is through diffusion and active transport through the mucous intestinal wall (Csáky, 1984). In humans, the surface area of the mucus cylinder with the same diameter and length as that of the small intestine is 0.33 m^2 (Carr & Toner, 1984; Kararli, 1995). However, since diffusion of nutrients is slow and transport is limited by distance, in order for the small intestine to efficiently absorb the nutrients from food, a large number of hierarchical folds such as valvulae conniventes, villi, and microvilli have developed to increase the surface area over 100-fold from the tubular cylindrical area of 0.33 m^2 to over 120 m^2 (Kararli, 1995)—some even estimate it to be as much as 200 m^2 (Wilson, 1962). The largest size scale of folding is the coiling of the entire organ itself, allowing 6 m of small intestine to fit snugly inside the body. The largest, macroscopic folds within the intestine are the valvulae conniventes which are circular rings of folds within the intestine, unevenly distributed in the small intestine. Their main function is to help the jejunum absorb nutrients by increasing surface area and to slow intestinal flow for better absorption (Abrahams, 2007; Quain, Schafer, & Thane, 1896).

Upon and between the valvulae conniventes, the small intestine is patterned with many protuberous folds of mucosal tissue known as villi, as well as deep narrow clefts known as crypts (Carr & Toner, 1984; Quain et al., 1896). The numerous villi (Fig. 9.2B), which slow intestinal flow, help increase absorption and overcome some of the distance limitations of transport by extending into the lumen of the intestine (Csáky, 1984). Each villus has a broad base and tapers near the tip, averaging about 0.1–0.25 mm in diameter (Marsh & Swift, 1969), and 0.5–0.8 mm in length (Quain et al., 1896). The surface of each villus is grooved and corrugated on an even smaller scale which further increases the surface area (Fig. 9.2C). The increased surface area facilitated by the grooves in the villi allows more numbers of columnar epithelial cells to line the surface, thereby increasing absorption. Upon closer examination, at the cellular level, the columnar cells that line the intestine have microvilli on their apical surface—additional tiny folds that even further increase the surface area at a smaller length scale (Fig. 9.2D). The cumulative effect of these patterns of folds upon folds upon folds is what allows for the great increase in absorption and other functions that would be significantly less efficient in a smooth, cylindrical mucus layer.

9.2 SYNTHETIC METHODS TO PATTERN CURVED AND FOLDED TISSUES

9.2.1 **Top-down synthetic approaches for curved and folded micropatterns**

Several top-down methods for synthesis of curved and folded micropatterns have been implemented with some success. Photolithographic methods borrowed from the electronics industry have enabled high-precision (<500 nm) patterning of features but primarily in two dimensions (Griffith & Swartz, 2006). A related approach, namely soft lithography involves patterning with soft substrates such as polymers or gels and is especially relevant to cell culture and tissue engineering. Many soft-lithographic approaches utilize an organic, flexible silicon-based polymer called poly(dimethylsiloxane) (PDMS). A number of soft-lithographic approaches have been developed to reproduce the curved and patterned micropatterns of biology. For an excellent review on soft lithographical methods, see Whitesides, Ostuni, Takayama, Jiang, and Ingber (2001), Andersson and van den Berg (2004), Gates, Xu, Love, Wolfe, and Whitesides (2004), Qian and Wang (2010), and Qin, Xia, and Whitesides (2010).

As illustrated in Fig. 9.3, soft lithography methods can be used to approximate curved and folded micropatterns with biocompatible materials such as polyurethanes (Fig. 9.3A–C) (Xia, Kim, & Whitesides, 1996). The approach has also been utilized to develop platforms to shape single cells (Fig. 9.3D) (Takeuchi, DiLuzio, Weibel, & Whitesides, 2005) and tissues (Fig. 9.3E and F) (Shin et al., 2004). For example, PDMS can be used to emboss patterns of biomaterials (replica molding) or inked with peptides, proteins, or cell suspensions and used to imprint these patterns on other surfaces (microcontact printing). Microcontact printing with selective biomolecules can then be used to create cell-adherent or cell-non-adherent regions (Whitesides et al., 2001). Then, the topographically patterned PDMS master can be inverted against a substrate to form a mold. Liquid polymer precursors can be injected into the cavities in the mold and subsequently cross-linked or cured (microtransfer molding). Furthermore, if the patterns on the slab of PDMS are continuous channels, the slab can be inverted on a glass slide to form capillary channels (microfluidics), and these channels can be filled with other kinds of prepolymers and cured (micromolding in capillaries or MIMIC).

Another related approach is to generate building blocks, such as films, beads, and fibers using microfluidic approaches and subsequently positioning these components in hierarchical curved and folded patterns. An illustrative example is the synthesis and use of monodisperse and micropatterned beads in cell culture (Matsunaga, Morimoto, & Takeuchi, 2011). In this example, cells were cultured on collagen beads and the cell-bead conjugates were used as building blocks for the assembly of macroscopic structures. These building blocks were synthesized via an asymmetric flow-focusing device and then impregnated by cells and introduced into a doll-shaped PDMS mold (Fig. 9.4A). The authors argue that the manufacture of this larger molded structure demonstrates the potential for tissue synthesis and rapid prototyping in geometries specific to patient needs, which could be derived from tomographic or magnetic resonance imaging (Lin, Kikuchi, & Hollister, 2004; Yeong, Chua, Leong, & Chandrasekaran, 2004). Simultaneously, the compatibility of the beads with multiple cell lines opens the doors to patterned co-culture of different cell types at a resolution better than 300 µm. In another study, sheets of microporous cell scaffolds were patterned with topographic features to support cell attachment and alignment (Papenburg et al., 2009). The scaffold layers were then manually assembled into well-stacked and tube-stacked structures manually (Fig. 9.4B). While the curved

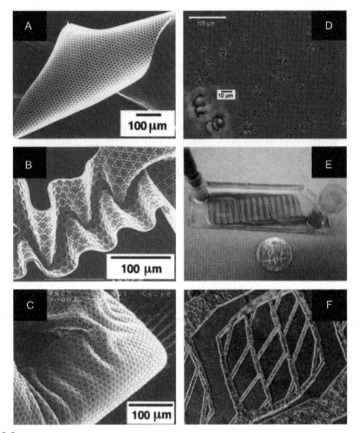

FIGURE 9.3

Soft lithography methods that approximate curved and folded micropatterns. (A–C) SEM images of patterned polyurethane microstructures that were formed using MIMIC, displayed on films of Saran Wrap. These patterned films folded into different shapes; the PU microstructures adhered well and roughly retained their forms. The strips a and b were produced accidentally on certain regions of the samples when folding the films mechanically. Adapted with permission from Xia et al. (1996), © 1996, American Chemical Society. (D) Phase-contrast microscopy image of spiral, filamentous cells against a background of agarose microchambers; the image inset shows spiral cells at a higher magnification. The microchambers used in this experiment had the following dimensions: diameter of the inner circle = 4 μm diameter of the outer circle = 13 μm, and height = 2 μm. Adapted with permission from Takeuchi et al. (2005), © 2005, American Chemical Society. (E, F) Macroscopic view of the microfabricated PDMS network. The patterned PDMS has been bonded to a flat PDMS sheet to obtain closed channels. The channels have been filled with ink for visualization. The network has a surface area of 13.69 cm^2 and an internal volume of 21.3 mm^3, with capillaries width in 5 mm (main channels near inlet and outlet) to 35 μm (capillary bed in the center) range. All channels have a depth of 35 μm. (F) HMEC-1 cells were maintained in culture for 14 days. Original magnification 50×. Panels E, F are reprinted from Shin et al. (2004) with kind permission from Springer Science + Business Media, Copyright © 2004 Springer Science + Business Media. (See color plate.)

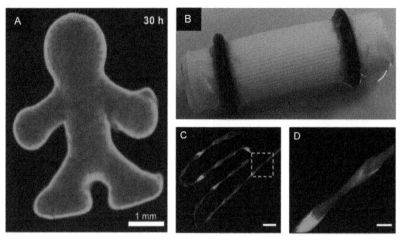

FIGURE 9.4

Modular methods of assembling micropatterns. (A) Macroscopic image of the 3D architecture constructed by molding of monodisperse cell beads imaged using the live/dead assay kit. Reproduced with permission from Matsunaga et al. (2011), Copyright © 2011 Wiley-VCH Verlag GmbH & Co. KGaA, Weinheim. (B) Illustrations of multilayer stacking of the phase separation micromolding sheets. Reprinted with permission from Papenburg et al. (2009), © 2009, Elsevier. (C) Fluorescence micrograph and (D) magnification showing the area outlined in yellow in (C). (See color plate.)

Reprinted with permission from Kang et al. (2011), © 2011, Macmillan Publishers Ltd.

microtopography of the layers enabled alignment of cells within the layers, stacking and rolling allowed the expression of curvature on a different length scale.

In addition to beads and layers, recently the digitally controlled microfluidic fabrication of fibers inspired by spider silk spinning process has been used to create spatially controlled co-cultures of cells for tissue engineering applications. In this example, microfluidic channels, set up in parallel, shoot out fibers of different composition to form spindles (Kang et al., 2011). To explore the implications and utility of this design in tissue engineering, hepatocyte and fibroblast cell cultures were selected as feeder solutions. Upon spinning, the cells were found to be viable 1 and 3 days later when encapsulated within the fiber structure. Significantly, the co-culture of the two types of cells showed enhanced cell survival compared to single cell-type fibers validating the utility of spun co-cultures for tissue engineering purposes. Furthermore, not only could different cell lines be embedded in any of the spindles, but the composition and the diameter of the spindles could be tuned across the length of the fiber and curved and folded morphologies could be formed (Fig. 9.4C and D). Related nozzle-based approaches such as 3D printing also provide an attractive means to shape tissues with curved and folded micropatterns with some limitations based on the linear rastering of the nozzle in the x, y, and z direction and the serial

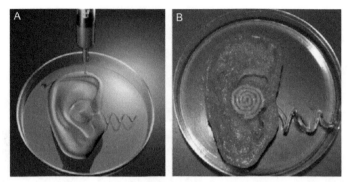

FIGURE 9.5

Three-dimensional printing of bionic ears with curved cartilaginous features and receiving antenna. (A) Schematic, and (B) printed ear. (See color plate.)

Reproduced with permission from Mannoor et al. (2013), © 2013, American Chemical Society.

nature of these processes. Nevertheless, a variety of tissue motifs have been patterned using this approach including bionic ears (Fig. 9.5) (Mannoor et al., 2013; Mironov et al., 2009).

9.2.2 Self-assembly of curved and folded micropatterns

The spontaneous formation of the neural tube from a flat sheet of neuroepithelial cells is a crucial event in the development of the central nervous system in humans (Wallingford, 2005). In fact, the differential strain generated by processes such as apical contraction in sheets of embryonic epithelial cells is a critical element in morphogenesis (Joshi & Davidson, 2012). Inspired by such biological processes, engineers have been attempting to shape curved and folded micropatterns by manipulating the strain in thin films so that they curve or fold without the need for any human intervention. These self-assembly paradigms, referred to as self-folding, are compatible with high-resolution planar lithographic approaches (Leong, Zarafshar, & Gracias, 2010; Ionov, 2011; Randall, Gultepe, & Gracias, 2012; Shenoy & Gracias, 2012). They can thus be used to transform precisely patterned sheets into curved and folded shapes with sizes ranging from centimeters to nanometers, in a highly parallel manner. In these approaches, thin film materials are deposited and layered, typically on a sacrificial layer. The thin films can be deposited by a variety of means including dip-coating, casting, nozzle printing, or spin coating. Films can be patterned using photopolymerization, nanoimprint lithography, electron beam lithography, or 3D printing so that any desired complexity of micropatterns can be formed. Differential strain can be generated by depositing and patterning heterogeneous films composed of materials of different cross-link density, chemical affinity, molecular weight, and stress. On dissolution of the sacrificial layer and immersion in solutions such as cell culture media, differential swelling causes the films to curve and fold on their own. Early work in this area utilized polyelectrolyte bilayers composed of polypyrrole doped with dodecylbenzenesulfonate and gold that folded in aqueous electrolytes due to swelling of the polymer on application of negative voltages as a result of cation insertion (Smela, Inganäs, & Lundström, 1995). One of the limitations of this approach was that folding and curving required an electrical wire to be connected to the structure; however, this limitation was overcome by utilizing folding mechanisms that relied on thermal- or chemical-affinity-derived strain gradients. For example, so-called bi-gel strips composed of the well-known temperature responsive gel, N-isopropylacrylamide and a polyacrylamide (PAAM) gel were used to form curved structures by modulating the temperature or changing the solvent (Hu, Zhang, & Li, 1995). However, these structures had very simple shapes and their compatibility with cell culture was not demonstrated.

In order to extend these approaches to tissue engineering, it is essential to utilize folding mechanisms that are compatible with living systems so that structures can fold and stay curved in culture media. A convenient approach is to utilize the residual stress generated during deposition of thin films such as chromium to drive curving and folding of scaffolds for cell culture (Jamal, Bassik, Cho, Randall, & Gracias, 2010). In this approach, precisely micropatterned self-assembling scaffolds were used to direct the assembly of fibroblasts in curved and folded geometries reminiscent of anatomically relevant geometries such as cylinders, spirals, and bidirectionally folded sheets. Interestingly, in these studies, filopodial connections were observed between cells on adjacent folds in the sheet which supports the hypothesis that folding morphologies may enhance cell–cell interactions. The presence of micropatterns within the sheet also enabled different regions of the scaffold to be cell adherent.

Other approaches including thermally responsive materials and surface forces have been utilized to create folded micropatterns for cell encapsulation therapy and drug delivery (Azam, Laflin, Jamal, Fernandes, & Gracias, 2011; Fernandes & Gracias, 2012; Stoychev, Puretskiy, & Ionov, 2011). Folded micropatterned capsules feature porosity in all three dimensions which is important to enhance mass transport of encapsulated cells with the surrounding medium which enhances cell viability and minimizes necrotic zones within the capsule (Randall, Kalinin, Jamal, Manohar, & Gracias, 2011). Three-dimensional micro and nanoporous capsules suggest an attractive design for bioreactors, artificial organs, and cell therapy. Porous folded capsules can also provide precise tunability for drug delivery and spatiotemporally controlled chemical release for a variety of biological applications (Kalinin, Murali & Gracias, 2012).

In addition to micropatterns for positioning cells in curved and folded geometries, one of the fundamental challenges in tissue engineering is the need to integrate fluidic channels and vasculature within these tissue constructs. One demonstrated approach is based on the generation of strain gradients by differential ultraviolet (UV) cross-linking in photoabsorbable films (Jamal, Zarafshar, & Gracias, 2011). Due to absorption, the intensity of the light is attenuated along the thickness of the thin film resulting in photo-cross-link gradients (CLGs). Such gradients were used to drive spontaneous curving and folding of polymers such as SU-8, which is a negative photoresist that is widely utilized in soft lithography and microfluidics. Since curving is generated by exposure to light, regions that are overexposed get strongly cross-linked and stay flat, while those exposed from the top curve downward and those exposed from the bottom curve upwards. Therefore, it is possible to form micropatterns in cylindrical, cubic, and bi-directionally folded geometries (Fig. 9.6). It was also possible to utilize this self-folding film as a kind of spine to curve and fold microfluidic networks of significant complexity (Fig. 9.7). These self-assembled networks enabled perfusion of chemicals within cell culture either on the scaffold or in its vicinity, and importantly in curved or folded geometries.

Recently, the self-assembly of curved and folded micropatterns with cell-laden hydrogels has also been demonstrated (Jamal et al., 2013). The approach utilized photoencapsulation of insulinoma and fibroblast cells within cross-linked

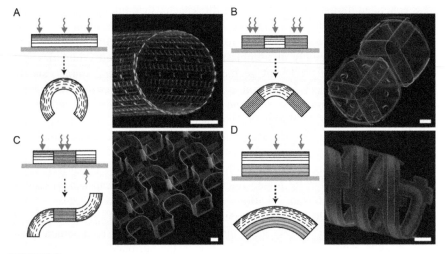

FIGURE 9.6

Self-curving and self-folding polymer patterns. (A–D) Schematics and fluorescence images of differentially photo-cross-linked and self-assembled SU-8 geometries. (A) A cylindrical mesh with hollow rectangular micropatterns and a uniform radius of curvature was self-assembled by creating a uniform cross-link gradient (CLG) across the entire SU-8 film. (B) Patterned and unpatterned cubes were self-assembled with CLG hinges and flat, high-cross-linked square faces. (C) An SU-8 sheet exhibiting bidirectional curvature was self-assembled with high-cross-linked square faces and both top and bottom exposed CLG hinges. (D) A hybrid SU-8/PDMS microfluidic device was self-assembled using a CLG-containing SU-8 layer to curve an underlying PDMS channel. Red lines indicate a segment of the hollow microfluidic channel. (See color plate.)

Reprinted with permission from Jamal et al. (2011), © 2011, Macmillan Publishers Ltd.

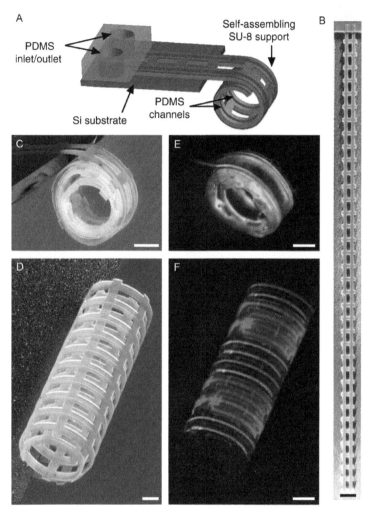

FIGURE 9.7

Self-assembling microfluidic devices. (A) An illustration of a self-assembling microfluidic device with PDMS inlets/outlets attached to a Si substrate and with PDMS channels integrated with a differentially cross-linked SU-8 film. (B, C) A brightfield image of a 3.5-cm long-multilayer SU-8/PDMS microfluidic device containing a single channel, (B) as patterned on a Si substrate, and (C) after self-assembly. (D) A brightfield image of a self-assembled microfluidic device with dual channels. (E, F) Fluorescence images showing the flow of (E) fluorescein (green), and (F) fluorescein (green)/rhodamine B (red) through single and dual channel devices, respectively. Scale bars are 500 μm (1 mm for B). (See color plate.)

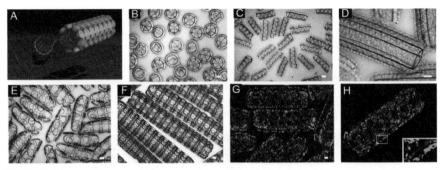

FIGURE 9.8

Origami inspired self-assembly of cell-laden hydrogels. (A) Conceptual schematic illustrating the self-folding of an initially planar PEG bilayer that contains different populations of cells in the inner (blue) and outer (green) hydrogel layers. (B–G) Examples of self-folded PEG hydrogels achieved: (B) spherical capsules, (C) helices, (D) cylindrical hydrogels with microtopographical posts on their surfaces, (E) small and (F) large cylindrical hydrogels with micropatterned holes, (G) cell-laden cylindrical bio-origami hydrogels containing micropatterned holes and photoencapsulated fibroblasts labeled with the viability stain calcein AM (green). Scale bars are 200 μm. (H) Multiculture of cells in distinct layers of a self-folded hydrogel. A deconvoluted fluorescent micrograph of a self-folded bilayer containing Hoechst-stained fibroblasts (blue) in the inner hydrogel layer and calcein AM-stained fibroblasts (green) in the outer hydrogel layer. (See color plate.)

Reprinted with permission from Jamal et al. (2013), © Copyright © 2013 Wiley-VCH Verlag
GmbH & Co. KGaA, Weinheim.

polyethyleneglycol diacrylate (PEG-DA) gels. Differential strain was achieved by differential swelling of two layers with different molecular weights and consequently different stiffness and swelling ratios. The process is versatile and a variety of micropatterned cell-laden hydrogels could be formed including those with multiple cell types (Fig. 9.8). Interestingly, it was also found that the insulin levels measured from insulinoma cells cultured in curved micropatterned geometries were higher than that in flat geometries which could be due to increased cell growth or cell–cell interactions in curved geometries. Elsewhere, cell traction forces generated by fibroblasts, smooth muscles, and endothelial cells were also utilized to self-fold parylene micropatterns (Kuribayashi-Shigetomi, Onoe, & Takeuchi, 2012).

9.3 PROTOCOL FOR SELF-ASSEMBLY OF CURVED AND FOLDED HYDROGELS

In this section, we describe the conceptual protocol to create curved and folded micropatterns in hydrogels using self-assembly. More details of specific experiments are available in published papers such as Azam et al. (2011) and Jamal et al.

(2011, 2013). The first step is to micropattern thin films of polymers or hydrogels in two dimensions using molding or photopatterning. To do so, one typically first drop casts or spin coats a solution containing the prepolymer such as PEG-DA (Aldrich) and UV polymerization initiator such as Irgacure (CIBA). For certain hydrogels or polymers such as PAAM, cross-linkers such as BIS-Acrylamide (Aldrich) can also be added. After depositing a thin film, portions of the film are exposed to UV light through transparent regions in a photomask so that those regions get cross-linked while light is blocked in opaque regions of the photomask. Any desired patterns can be designed using vector graphics programs and these designs can be printed on transparency films using high resolution printers or sent to commercial vendors for photomasks.

The important point to note is that a film will curve or fold only when (a) it is released from the substrate which is typically achieved by depositing a sacrificial layer and (b) there is a bending moment. A wide range of sacrificial layers have been utilized and one needs to ensure that dissolution of the sacrificial layer can be achieved without dissolution of the curving or folding micropatterned film. Examples include metals such as copper which can be dissolved in acids or commercial etchants and polyvinyl alcohol (dissolved in water). To generate a bending moment strain can be created in the films using bilayers with different swelling characteristics (Bassik, Abebe, Laflin, & Gracias, 2010), depositing a low melting point polymer to enable strain on melting by surface forces (Azam et al., 2011), creating a CLG in the thin film (Jamal et al., 2011). Other examples include creating photo-cross-link lateral heterogeneities or utilizing stimuli responsive films and recent review articles (Fernandes & Gracias, 2012; Gracias, 2013) survey these methods. The important parameters are the magnitude of stress, the thickness of the films, and their Poisson ratio which ultimately determines the tightness or radius of curvature of the final structures that result. There are a variety of models (such as one described in Jamal et al., 2013) that can be used to predict radii of curvature of simple geometries such as rectangular beams while more extensive finite element modeling may be required for more complicated geometries. Importantly, in order for the structures to be applicable in cell culture applications they must remain curved or folded and not flatten out in biological media.

SUMMARY

In summary, just like in the human body, engineers must learn to position and pattern cells, fluidic channels, and biomolecules in curved and folded geometries so as to create anatomically realistic *in vitro* models to study biological processes and investigate disease. In order to do this, it is necessary to learn how to create cell-adherent, microfluidic, and cellular micropatterns in curved and folded geometries. The mastery of such approaches could also lead to the development of synthetic replacement organs. Significant progress has been made since the days of cell culture on Petri dishes, but challenges abound especially in the creation of anatomically complete systems with precise hierarchy from the cellular upward to the tissue and organ level. Further, the ability to observe and mimic biological morphogenesis could enable revolutionary advances in the mass production of such micropatterned organoids and organs.

Acknowledgment

We acknowledge support from the National Science Foundation grant NSF CBET-1066898.

References

Abrahams, P. (2007). *How the body works (a comprehensive illustrated encyclopedia of anatomy)*. London: Amber Books Ltd.

Andersson, H., & van den Berg, A. (2004). Microfabrication and microfluidics for tissue engineering: State of the art and future opportunities. *Lab on a Chip, 4*(2), 98–103.

Arakawa, M. (1971). A scanning electron microscope study of the human glomerulus. *The American Journal of Pathology, 64*(2), 457.

Azam, A., Laflin, K. E., Jamal, M., Fernandes, R., & Gracias, D. H. (2011). Self-folding micropatterned polymeric containers. *Biomedical Microdevices, 13*(1), 51–58.

Bassik, N., Abebe, B., Laflin, K., & Gracias, D. H. (2010). Photolithographically patterned smart hydrogel based bilayer actuators. *Polymer, 51,* 6093–6098.

Berger, S. A., Goldsmith, E. W., & Lewis, E. R. (1996). *Introduction to bioengineering.* Oxford, UK: Oxford University Press.

Bergstrand, A., & Bucht, H. (1958). Anatomy of the glomerulus as observed in biopsy material from young and healthy human subjects. *Zeitschrift für Zellforschung und Mikroskopische Anatomie, 48*(1), 51–73.

Carr, K. E., & Toner, P. G. (1984). Morphology of the intestinal mucosa. In T. Z. Csáky (Ed.), *Pharmacology of intestinal permeation I* (pp. 1–50). Berlin, Heidelberg: Springer.

Csáky, T. Z. (1984). Intestinal permeation and permeability: An overview. In T. Z. Csáky (Ed.), *Pharmacology of intestinal permeation I* (pp. 1–50). Berlin, Heidelberg: Springer.

Dobson, C. M. (2003). Protein folding and misfolding. *Nature, 426*(6968), 884–890.

Eiraku, M., Takata, N., Ishibashi, H., Kawada, M., Sakakura, E., Okuda, S., et al. (2011). Self-organizing optic-cup morphogenesis in three-dimensional culture. *Nature, 472*(7341), 51–56.

Fernandes, R., & Gracias, D. H. (2012). Self-folding polymeric containers for encapsulation and delivery of drugs. *Advanced Drug Delivery Reviews, 64*(14), 1579–1589.

Gates, B. D., Xu, Q. B., Love, J. C., Wolfe, D. B., & Whitesides, G. M. (2004). Unconventional nanofabrication. *Annual Review of Materials Research, 34,* 339–372.

Gracias, D. H. (2013). Stimuli responsive self-folding using thin polymer films. *Current Opinion in Chemical Engineering, 2,* 112–119.

Griffith, L. G., & Swartz, M. A. (2006). Capturing complex 3D tissue physiology in vitro. *Nature Reviews Molecular Cell Biology, 7*(3), 211–224.

Hall, J. E. (2010). *Guyton and hall textbook of medical physiology: Enhanced E-book.* Elsevier Health Sciences: Saunders.

Hartl, F. U. (1996). Molecular chaperones in cellular protein folding. *Nature*, *381*(6583), 571–580.

Hegyi, H., & Gerstein, M. (1999). The relationship between protein structure and function: A comprehensive survey with application to the yeast genome. *Journal of Molecular Biology*, *288*(1), 147–164.

Helal, I., Fick-Brosnahan, G. M., Reed-Gitomer, B., & Schrier, R. W. (2012). Glomerular hyperfiltration: Definitions, mechanisms and clinical implications. *Nature Reviews Nephrology*, *8*(5), 293–300.

Herculano-Houzel, S. (2012). The remarkable, yet not extraordinary, human brain as a scaled-up primate brain and its associated cost. *Proceedings of the National Academy of Sciences of the United States of America*, *109*(Suppl. 1), 10661–10668.

Höhne, M., Ising, C., Hagmann, H., Völker, L. A., Brähler, S., Schermer, B., et al. (2012). Light microscopic visualization of podocyte ultrastructure demonstrates oscillating glomerular contractions. *The American Journal of Pathology*, *182*(2), 332–338.

Hu, Z. B., Zhang, X. M., & Li, Y. (1995). Synthesis and application of modulated polymer gels. *Science*, *269*(5223), 525–527.

Ionov, L. (2011). Soft microorigami: Self-folding polymer films. *Soft Matter*, *7*, 6786–6791.

Jamal, M., Bassik, N., Cho, J. H., Randall, C. L., & Gracias, D. H. (2010). Directed growth of fibroblasts into three dimensional micropatterned geometries via self-assembling scaffolds. *Biomaterials*, *31*(7), 1683–1690.

Jamal, M., Kadam, S. S., Xiao, R., Jivan, F., Onn, T.-M., Fernandes, R., et al. (2013). Bio-origami hydrogel scaffolds composed of photocrosslinked PEG bilayers. *Advanced Healthcare Materials*, *2*(8), 1142–1150.

Jamal, M., Zarafshar, A. M., & Gracias, D. H. (2011). Differentially photo-crosslinked polymers enable self-assembling microfluidics. *Nature Communications*, *2*, 527.

Joshi, S. D., & Davidson, L. A. (2012). Epithelial machines of morphogenesis and their potential application in organ assembly and tissue engineering. *Biomechanics and Modeling in Mechanobiology*, *11*(8), 1109–1121.

Kalinin, Y. V., Murali, A., & Gracias, D. H. (2012). Chemistry with spatial control using particles and streams. *RSC Advances*, *2*(26), 9707–9726.

Kang, E., Jeong, G. S., Choi, Y. Y., Lee, K. H., Khademhosseini, A., & Lee, S. H. (2011). Digitally tunable physicochemical coding of material composition and topography in continuous microfibres. *Nature Materials*, *10*(11), 877–883.

Kararli, T. T. (1995). Comparison of the gastrointestinal anatomy, physiology, and biochemistry of humans and commonly used laboratory animals. *Biopharmaceutics and Drug Disposition*, *16*(5), 351–380.

Kuribayashi-Shigetomi, K., Onoe, H., & Takeuchi, S. (2012). Cell origami: Self-folding of three-dimensional cell-laden microstructures driven by cell traction force. *PLoS One*, *7*(12), e51085.

Leong, T. G., Zarafshar, A., & Gracias, D. H. (2010). Three dimensional fabrication at small size scales. *Small*, *6*(7), 792–806.

Lin, C. Y., Kikuchi, N., & Hollister, S. J. (2004). A novel method for biomaterial scaffold internal architecture design to match bone elastic properties with desired porosity. *Journal of Biomechanics*, *37*(5), 623–636.

Mannoor, M. S., Jiang, Z., James, T., Kong, Y. L., Malatesta, K. A., Soboyejo, W. O., et al. (2013). 3D printed bionic ears. *Nano Letters*, *13*(6), 2634–2639.

Marsh, M. N., & Swift, J. A. (1969). A study of the small intestinal mucosa using the scanning electron microscope. *Gut*, *10*(11), 940.

Matsunaga, Y. T., Morimoto, Y., & Takeuchi, S. (2011). Molding cell beads for rapid construction of macroscopic 3D tissue architecture. *Advanced Materials*, *23*(12), H90–H94.

Mironov, V., Visconti, R. P., Kasyanov, V., Forgacs, G., Drake, C. J., & Markwald, R. R. (2009). Organ printing: Tissue spheroids as building blocks. *Biomaterials*, *30*(12), 2164–2174.

Moriya, T., Tanaka, K., & Moriya, R. (2000). Glomerular structural changes and structural-functional relationships at early stage of diabetic nephropathy in Japanese type 2 diabetic patients. *Medical Electron Microscopy*, *33*(3), 115–122.

Mundel, P., & Kriz, W. (1995). Structure and function of podocytes: An update. *Anatomy and Embryology*, *192*(5), 385–397.

Nørrelund, H., Christensen, K. L., Samani, N. J., Kimber, P., Mulvany, M. J., & Korsgaard, N. (1994). Early narrowed afferent arteriole is a contributor to the development of hypertension. *Hypertension*, *24*(3), 301–308.

Nyengaard, J. R., & Bendtsen, T. F. (1992). Glomerular number and size in relation to age, kidney weight, and body surface in normal man. *The Anatomical Record*, *232*(2), 194–201.

Papenburg, B. J., Liu, J., Higuera, G. A., Barradas, A. M. C., de Boer, J., van Blitterswijk, C. A., et al. (2009). Development and analysis of multi-layer scaffolds for tissue engineering. *Biomaterials*, *30*(31), 6228–6239.

Qian, T. C., & Wang, Y. X. (2010). Micro/nano-fabrication technologies for cell biology. *Medical and Biological Engineering and Computing*, *48*(10), 1023–1032.

Qin, D., Xia, Y., & Whitesides, G. M. (2010). Soft lithography for micro- and nanoscale patterning. *Nature Protocols*, *5*(3), 491–502.

Quain, J., Schafer, E. A., & Thane, G. D. (1896). *Quain's elements of anatomy*. London: Longmans, Green, and Co..

Randall, C. L., Gultepe, E., & Gracias, D. H. (2012). Self-folding devices and materials for biomedical applications. *Trends in Biotechnology*, *30*(3), 138–146.

Randall, C. L., Kalinin, Y. V., Jamal, M., Manohar, T., & Gracias, D. H. (2011). Three-dimensional microwell arrays for cell culture. *Lab on a Chip*, *11*(1), 127–131.

Rasch, R., Lauszus, F., Thomsen, J. S., & Flyvbjerg, A. (2005). Glomerular structural changes in pregnant, diabetic, and pregnant-diabetic rats. *APMIS*, *113*(7–8), 465–472.

Rockland, K. S. (2002). Non-uniformity of extrinsic connections and columnar organization. *Journal of Neurocytology*, *31*(3–5), 247–253.

Sasson, A. N., & Cherney, D. Z. I. (2012). Renal hyperfiltration related to diabetes mellitus and obesity in human disease. *World Journal of Diabetes*, *3*(1), 1–6.

Sato, T., & Clevers, H. (2013). Growing self-organizing mini-guts from a single intestinal stem cell: Mechanism and applications. *Science*, *340*(6137), 1190–1194.

Shenoy, V. B., & Gracias, D. H. (2012). Self-folding thin film materials: From nanopolyhedra to graphene origami. *MRS Bulletin*, *37*(9), 847–854.

Shin, M., Matsuda, K., Ishii, O., Terai, H., Kaazempur-Mofrad, M., Borenstein, J., et al. (2004). Endothelialized networks with a vascular geometry in microfabricated poly(dimethyl siloxane). *Biomedical Microdevices*, *6*(4), 269–278.

Shingleton, A. (2010). Allometry: The study of biological scaling. *Nature Education Knowledge*, *3*(10), 2.

Shirahama, T., & Cohen, A. S. (1967). Fine structure of the glomerulus in human and experimental renal amyloidosis. *The American Journal of Pathology*, *51*(5), 869–911.

Silverthorn, D. U. (2004). *Human physiology: An integrated approach*. San Francisco, CA: Pearson/Benjamin Cummings.

Smela, E., Inganäs, O., & Lundström, I. (1995). Controlled folding of micrometer-size structures. *Science, 268*(5218), 1735–1738.

Stoychev, G., Puretskiy, N., & Ionov, L. (2011). Self-folding all-polymer thermoresponsive microcapsules. *Soft Matter, 7*(7), 3277–3279.

Taber, L. A. (1995). Biomechanics of growth, remodeling, and morphogenesis. *Applied Mechanics Reviews, 48*, 487.

Takeuchi, S., DiLuzio, W. R., Weibel, D. B., & Whitesides, G. M. (2005). Controlling the shape of filamentous cells of *Escherichia coli*. *Nano Letters, 5*(9), 1819–1823.

Wallingford, J. B. (2005). Neural tube closure and neural tube defects: Studies in animal models reveal known knowns and known unknowns. *American Journal of Medical Genetics, 135C*(1), 59–68.

Whitesides, G. M., Ostuni, E., Takayama, S., Jiang, X., & Ingber, D. E. (2001). Soft lithography in biology and biochemistry. *Annual Review of Biomedical Engineering, 3*(1), 335–373.

Wickelgren, I. (1999). First components found for key kidney filter. *Science, 286*(5438), 225–226.

Wilson, T. H. (1962). *Intestinal absorption*. Philadelphia, PA: W. B. Saunders Co.

Xia, Y. N., Kim, E., & Whitesides, G. M. (1996). Micromolding of polymers in capillaries: Applications in microfabrication. *Chemistry of Materials, 8*(7), 1558–1567.

Yeong, W. Y., Chua, C. K., Leong, K. F., & Chandrasekaran, M. (2004). Rapid prototyping in tissue engineering: challenges and potential. *Trends in Biotechnology, 22*(12), 643–652.

Cell Migration in Confined Environments

10

Daniel Irimia

Massachusetts General Hospital, Harvard Medical School, and Shirners Hospitals for Children, Boston, Massachusetts, USA

CHAPTER OUTLINE

Abstract

We describe a protocol for measuring the speed of human neutrophils migrating through small channels, in conditions of mechanical confinement comparable to those experienced by neutrophils migrating through tissues. In such conditions, we find that neutrophils move persistently, at constant speed for tens of minutes, enabling precise measurements at single cells resolution, for large number of cells. The protocol relies on microfluidic devices with small channels in which a solution of chemoattractant and a suspension of isolated neutrophils are loaded in sequence. The migration of neutrophils can be observed for several hours, starting within

Methods in Cell Biology, Volume 121
ISSN 0091-679X
http://dx.doi.org/10.1016/B978-0-12-800281-0.00010-5

minutes after loading the neutrophils in the devices. The protocol is divided into four main steps: the fabrication of the microfluidic devices, the separation of neutrophils from whole blood, the preparation of the assay and cell loading, and the analysis of data. We discuss the practical steps for the implementation of the migration assays in biology labs, the adaptation of the protocols to various cell types, including cancer cells, and the supplementary device features required for precise measurements of directionality and persistence during migration.

INTRODUCTION

The standard textbook pictures of white blood cells moving through a homogenous space from circulation to a site of injury in tissues are often misleading. Far from moving in homogenous microenvironment, white blood cells encounter all sorts of obstacles during their journey through tissues *in vivo*. In particular, fast moving cells like neutrophils often have to squeeze in between other cells in the tissue, chimney through tiny capillaries, go around matrix fibers, and find passage ways through fibrin clots or scars (McDonald et al., 2010). Like the textbooks, most of the *in vitro* assays we are using today to study the migration of cells rarely acknowledge the complexity of the *in vivo* microenvironment. Traditional assays (Zigmond chamber, Dunn chamber, or micropipette assay) as well as the majority of microfluidic assays mostly observe the cells migrating on flat surfaces, without any of the tissue-relevant mechanical challenges. The limitations of the current assays are not just methodological, but they often preclude the decoupling of individual conditions and modulators of cell migration. One early example illustrating the new insights that could emerge from restoring the mechanical complexity of the cell migration microenvironment was the finding of calcium-independent cell migration after squeezing neutrophils in between glass and agarose gel (Malawista & de Boisfleury Chevance, 1997), a finding later confirmed also in dendritic cells (DCs) (Lammermann et al., 2008). More recently, emerging microfluidic platforms have taken the issue of mechanical complexity to a higher level of sophistication and increased precision of microenvironment control.

The use of microscale channels for mechanically confining the cells during migration creates opportunities for discovery and for designing more robust drug screening assays. For example, confining neutrophils to channels significantly smaller than the cell cross section has been shown to reduce the variability in speed during chemotaxis (Irimia, Charras, Agrawal, Mitchison, & Toner, 2007). In the absence of confinement, the variations in migration speed are a significant issue when analyzing neutrophil migration on flat surfaces. The reduced variability was important when analyzing the human neutrophil migration for the purpose of defining a normal range of values for neutrophil migration from healthy volunteers and for quantifying changes in patients (Butler et al., 2010). The confinement was also useful for measuring the migration of other leukocytes as well, as shown in studies using DCs (Faure-Andre et al., 2008; Renkawitz et al., 2009) and T cells (Jacobelli et al., 2010). More recently, precise comparisons of the migration speed and persistence of various cancer epithelial cells have been enabled by microfluidic devices that confined the moving cells to channels (Irimia & Toner, 2009; Scherber et al., 2012). In addition to the analysis of speed, small channels with bifurcations also helped quantify the directional decisions during migration in normal and cancer epithelial cells and in human neutrophils (Ambravaneswaran, Wong, Aranyosi, Toner, & Irimia, 2010; Scherber et al., 2012). In these devices, the directional decisions that cells make when encountering the bifurcations were quantified in binary mode, simplifying the analysis and comparisons between conditions. Additional challenges for the moving cells and opportunities for biological insights emerge from loading the channels with Matrigel (Wolfer et al., 2010), tapering the channels to small cross sections (Balzer et al., 2012; Gallego-Perez et al., 2012), or the combination of geometric and extracellular matrix conditions (Kraning-Rush, Carey, Lampi, & Reinhart-King, 2013). Applications are also emerging toward the identification of new drug targets for cell migration (Smolen et al., 2010) or new context for activities of existing compounds (Balzer et al., 2012; Rolli, Seufferlein, Kemkemer, & Spatz, 2010).

10.1 DESIGNING THE DEVICES

10.1.1 Size of the microchannels

The size and topography of the channels for cell migration is one important parameter that depends on the type of cells to be studied and the goals of the experiments. In the simplest design, straight channels, with cross section comparable to the size of the cells, enable precise measurements of cell speed. In the case of human neutrophils (10 µm average diameter for cells in suspension), channels having 6–8 µm width and 3 µm height appear to be optimal for observing robust cell migration. For the migration of cancer epithelial cells, larger channels (for example 10 µm × 10 µm) are most favorable (Irimia & Toner, 2009). The length of channels can vary from 100 to 1000 µm.

10.1.2 Cell loading chamber

Because the height of the migration channels is smaller than the cells in suspension, a second set of channels that are taller has to be used to accommodate the cells to the entrance of the migration channels. The cell loading chamber can be designed as tall as 100 µm. Cells introduced as a suspension in the devices, usually settle by gravity alone to the bottom of the cell loading chamber, closer to the entrance to the emigration channels.

10.1.3 **Gradient formation**

The chemokine gradient forms by diffusion between the cell migration channels and the cell loading chamber. The migration channels are designed as chemokine "sources", from which the cheomokine will diffuse at predictable rates, dependent on the length, geometry, and cross section of the channels. The cell loading chamber serves as a "sink" for the chemoattractant diffusing from the migration channels, the larger the chamber, the more effective it will be at maintaining the reduced chemokine concentration. To achieve the differential loading of the migration channels and cell loading chamber with chemokine and buffer, respectively, we are taking advantage of a sequential protocol for operating the devices. Off the shelf, the devices contain no liquid and are filled with air. In the first step, the entire device is primed with chemokine solution, which will fill the dead-end migration channels as well as the loading chamber. In the second step, the loading chamber is washed by flowing buffer from the inlet to the outlet. Because there is no flow in the dead-end migration channels, only the chemokine in the loading chamber is washed off. Most of the chemokine initially in the migration channels, remains in these channels. Quickly after, the diffusion of this chemokine between the "source" and the "sink" in the absence of convection produces the chemokine gradients which ultimately drive the cells into the side channels. Because of the limited volume of the source and sink, these gradients will progressively decrease and eventually the chemokine concentration becomes uniform. The time to equilibrium depends on the molecular weight of the chemokine and design parameters of the device. Longer channels, having smaller cross section toward the loading chambers will result in longer lasting gradients.

10.2 **DEVICE FABRICATION**

Fabricating the devices requires two major steps. In the first step, a mold will be fabricated, with the negative of the channels. In the second step, the mold will be replicated in elastomeric material and bonded to a glass slide, completing the channels. While the first step requires specialized photolithography techniques usually performed in clean room environment, the second step is relatively simple and could be accomplished in any lab. It is important to know that hundreds of devices could be fabricated from just one mold. Consequently, it is most economical for users interested in the application of the devices, to first fabricate one mold through the various microfluidic foundries, resource centers, or academic labs specialized in making microfluidic devices. The fabrication of the microfluidic devices could be accomplished locally, at the time when as they are needed for experiments. The essential steps for fabricating the elastomeric devices and final assembly are described in detail in this section.

10.2.1 **Materials**

a. Silicone elastomer (Polydimethylsiloxane (PDMS), Dow Corning Sylgard 184 Silicone Encapsulant Clear 0.5 kg Kit, Ellsworth Adhesives, Germantown, WI)

b. Vacuum desiccator (F42020-0000, Bel-Art Products, Wayne, NJ)
c. Vacuum oven (VO 914A, Thermo Scientific)
d. Hot plate
e. Scalpel and razor blades
f. Handling tweezers (item # 758TW070 TechniTool, Worcester, PA)
g. Punch—Harris Unicore, 1 mm (item #15072, Ted Pella)
h. Precleaned glass slides or glass coverslips
i. Five inch plastic petri dishes
j. Weigh dishes (item #01018-04 Cole Parmer, Court Vernon Hills, IL)
k. Plastic forks
l. Wipes

10.2.2 Equipment

a. Balance (EL2001, Mettler Toledo, Columbus, OH)
b. Plasma cleaner (PDC32 G, Harrick Plasma, Ithaca, NY)

10.2.3 Method

a. Secure the silicon wafer (the mold) to the bottom of the plastic dish with tape. Position the wafer to the center of the dish for easier cutting of the elastomer after curing and uniform thickness.
b. Prepare the uncured elastomer by mixing the base and curing agent in 10:1 ratio. Prepare about 50 g of elastomer the first time when using a mold. After cutting out the first set of devices, only 15 g of elastomer is routinely necessary. Place the mixing tray on the balance and zero the reading. Add first the curing agent which is more fluid to the center of the tray. Then, add the base, which is more viscous, around the drop of curing agent. For a routine preparation, you would need 1.5 g of curing agent and 15 g of the base. Use the fork to mix the base and curing agent thoroughly for at least 2 min. Air bubble will form upon vigorous mixing and the mixture should look white at the end of mixing.
c. Pour the uncured elastomer over the wafer making sure the coverage is uniform. Place the dish in the vacuum jar for 30–60 min. The elastomer should look clear and transparent.
d. Place the dish in the oven set to 65 °C for 8–12 h. Make sure the shelves in the oven are leveled; this is very important for the uniform thickness of the devices.
e. Cut the elastomer off the surface of the mold. Using the scalpel, cut close to the edge of the mold, on the surface of the silicon wafer. Using clean gloves and tweezers, peel off slowly the cured elastomer from the surface of the mold, with extra care where the smaller features of the devices are. Place the elastomer on the flat, clean surface of a cutting board, with the features down. It is advisable that you cover the mold with a fresh layer of uncured elastomer, preventing dust or dirt from contaminating the surfaces. The mold with cured elastomer could be stored indefinitely at room temperatures.

   f. From the piece of elastomer, cut smaller pieces to the size of the devices, using the razor blades. Punch the inlet and outlet holes using the core punch.

   g. Place the glass slide (or coverslip) and the smaller piece of elastomer with the features facing up, inside the plasma machine. Follow the step-by-step instructions from the plasma machine manufacturer. You will notice the formation of purple plasma inside the plasma chamber. Expose the glass and elastomer surfaces to plasma for approximately 30 s. It is important to optimize the time of exposure to plasma for strongest bonding, depending on the particularities of the machine being used. Remove the glass slide and elastomer piece from the plasma machine using tweezers, careful not to touch the surfaces with your fingers.

   h. Turn the elastomer upside down on the glass slide and press gently. Looking carefully at an approximately 45° angle you will notice the change in reflection at the interface between the two materials, indicating the effective bonding of the two surfaces. It is very important to know that you have only one chance of bringing the two surfaces together. If the two elastomer pieces need to be repositioned on the glass surface after the two surfaces come into contact, the plasma step has to be repeated.

   i. Finally, place the device on the hotplate set to 90 °C for 3 min for extra strong bonding.

10.3 MEASURING CELL MIGRATION
10.3.1 Materials
- *Regents*
  - Human fibronectin (Sigma Aldrich, St. Louis, MO)
  - Hank's buffered salt solution (Life Technologies, Grand Island, NY)
  - Food dyes (for training purpose only)
- *Materials*
  - 1 ml syringes (item # 309659, Becton Dickinson, Franklin Lakes, NJ)
  - Blunt syringe needles (BN3005, 30 G × ½ in., Brico Medical Supplies, Dayton, NJ)
  - Tygon tubing ID=0.01 in., OD=0.03 in. (S54HL Tygon Tubing, Greene Rubber, Woburn, MA)
  - Hemostats forceps (item # 13-812-45, Fisher Sci, Pittsburgh, PA)
  - Flat tip tweezers (item # 758TW462, TechniTool)

10.3.2 Equipment
- Microscope and camera with time-lapse capabilities (Nikon TiE, Nikon, Japan)
- Environmental chamber (Live Cell, Pathology Devices, Westminster, MD)
- Image J software

10.3.3 **Method**

1. Prepare four 4 cm long pieces of the tubing for each channel of the device (Fig. 10.1). Fit 30 G blunt needles to one end of three of these pieces of tubing. For this, gently slide the tubing over the needle tip using your fingers.
2. Insert one piece of free tubing and one with a needle into the inlet and outlet holes of device using tweezers.
3. Fill one 1 ml syringe with 300 µl of the chemokine solution, connect the syringe to the needle, and slowly push the solution into the device. As soon as the solution starts coming out of the free end of the tubing at the outlet of the device, clamp that tubing using the forceps.
4. Continue pushing the syringe plunger. This will push the fluid into the side channels of the device (Fig. 10.2). Allow enough time for the air trapped in the dead-end side channels to escape by diffusion through the elastomer and into the chemokine solution. You could verify on a microscope that all the air has been removed from the side channels and the entire device is filled with chemokine solution. Remove the tubing with the syringe and leave one droplet of chemokine solution on top of the inlet hole.
5. Prepare a second syringe with 300 µl of buffer solution, connect it to a new piece of tubing, and gently fill the tubing with buffer until a droplet forms at the free end of the tubing. Using the tweezers, insert this tubing into the inlet hole of the device. Open the clamp on the outlet tubing and gently push the buffer through the device for approximately 10 s (Fig. 10.3). This step is very important,

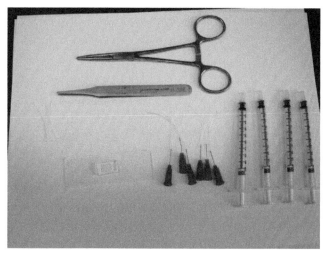

FIGURE 10.1

Microfluidic device on a standard glass slide, tubing, syringes, tweezers, and forceps necessary for setting up the cell migration experiments. (For color version of this figure, the reader is referred to the online version of this chapter.)

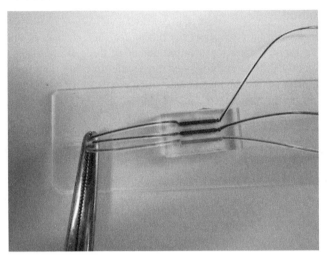

FIGURE 10.2

The microfluidic device primed with food dyes. Notice the array of side channels which are also filled. (For color version of this figure, the reader is referred to the online version of this chapter.)

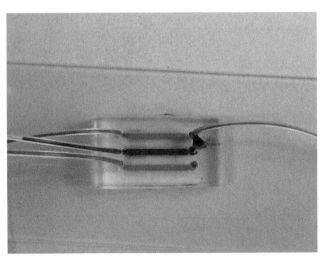

FIGURE 10.3

Soon after replacing the solution in the top "cell loading channel," a gradient begins to form along the side channels. The droplet of fluid on top of the inlet (right) has to be cleaned before the tubing is removed. (For color version of this figure, the reader is referred to the online version of this chapter.)

as it will remove the chemokine from the main channel and trigger the formation of the chemokine gradient between the main channel and the end of the side channels (Fig. 10.4). Clean the inlet and outlet drops of fluid. Clamp the outlet tubing and gently remove the inlet tubing and syringe, leaving a droplet of buffer solution on top of the inlet hole.

6. To introduce the cells, fill the cell suspension in the third syringe at density of 10^6 cells/ml or higher. Connect it to a new piece of tubing, and gently fill the tubing with the cell suspension until a droplet forms at the free end of the tubing. Using the tweezers, insert this tubing into the inlet hole of the device. Open the clamp on the outlet tubing and gently push the cell suspension through the device for approximately 3 s. Verify on a microscope that enough cells are present in the cell loading channel. If needed, flow more cells into the device, keeping in mind that the cells will settle inside the syringe with time, affecting the local cell density in the tubing. Remove tubing at inlet and close the channel by looping tubing in outlet over to the inlet.

7. Once the cells are loaded in the device, place the device in the environmental chamber on the microscope and start recording images (Fig. 10.5). The magnification and timing between images will depend on the size of the cells and purpose of the experiments. If measuring the speed of migration is the goal, $10\times$ magnification using phase contrast and 20-s interval between frames is

FIGURE 10.4

Spatial gradient of fluorescence along the cell migration channels. The cell loading chamber on the left also shows up in fluorescence because of the order of magnitude increased thickness. (For color version of this figure, the reader is referred to the online version of this chapter.)

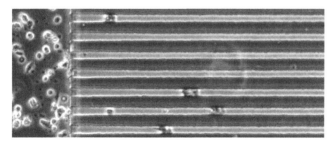

FIGURE 10.5

Human neutrophils from a healthy volunteer migrating through the channels in response to fMLP gradients (formyl-Methionyl-Leucyl-Phenylalanine, 100 nM maximum concentration), at 17 min after loading in the device.

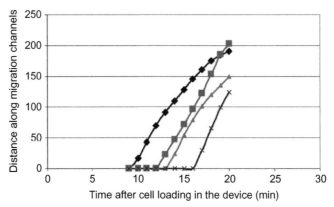

FIGURE 10.6

Example of neutrophil displacement versus time plots quantifying the migration in response to fMLP gradients. The four neutrophils enter difference channels at different times after cell loading but move at comparable and constant speed. (For color version of this figure, the reader is referred to the online version of this chapter.)

usually sufficient. If the microscope has an automated stage, images from multiple locations in the device could be acquired in parallel stacks.

8. Images can be analyzed manually using the manual tracking routine in Image J http://rsb.info.nih.gov/ij/plugins/track/track.html. You could then export the data to Excel or other programs that could help with the presentation of the data and its analysis (Fig. 10.6). Options for automated tracking also exist, in particular if cells are fluorescently labeled before the migration assay. While many of the fluorescent dyes for cell tracking can interfere with the migration, it is important to compare the speed of cells tracked manually in phase-contrast images to the speed of labeled cells in fluorescence images.

10.4 DISCUSSION

Successful cell migration assays imply reproducible results from the same sample, independent of the operator. In our experience working with various microfluidic devices for cell migration, the confinement of cells in channels provides the most reliable devices, robust to perturbations and immune to large number of variables that one could encounter in experimental work. The learning curve for someone who has not used microfluidic devices before could be rather steep. However, after the initial setup hurdles fabricating and using microfluidic devices could become routine and the experimenter could focus quickly on the application rather than the technology. The ability to visualize the cells in channels at the different stages of the protocol provides immediate feedback which is not usually available through any other methods.

The migration assay could be easily adapted for probing the chemotaxis of cancer cells in gradients of growth factors (Desai, Bhatia, Toner, & Irimia, 2013). To accommodate for the larger cell size, the cross section of the channels is increased to 10×10 µm. Cancer cells can move through smaller as well as larger channels, but their migration speed decreases significantly in either one of the situations (Irimia et al., 2009). While cancer cells move an order of magnitude slower than neutrophils, the side channels also have to accommodate the need for gradients of growth factors that last for more than 12 h. This can be accomplished by extending the length of the side channels to increase the distance for diffusion or by designing larger reservoirs at the end of the side channels to increase the amount of factors available (Jones et al., 2012).

The protocol presented in this chapter is robust and adaptable to various purposes. For example, by implementing post and bifurcations in the design of the channels, the directionality of the moving cells could be measured directly, from the binary decisions at these locations (Ambravaneswaran et al., 2010; Scherber et al., 2012). Moreover, the design of migration channels can be easily modified to accommodate various cell types and various cell sizes. The length and geometry of the migration channels could be modified to increase the duration of the gradient for slower moving cells.

10.5 **PRACTICAL NOTES**

- Cut the tubing at an angle to facilitate inserting the tubing into the holes of the elastomeric devices.
- To avoid bubbles, pipette small amount of phosphate buffered saline over inlet tubing such that a droplet of liquid surrounds the tubing before removing the chemokine-filled tubing from the inlet.
- You should practice setting up the device using concentrated food dyes in water, without cells. The concentrated dye should allow you to see the gradient formed along the channels after flushing the main channel with water, providing quick feedback on the technique.
- The first image of the device before cells enter the channels could be used as reference and subtracted from subsequent images to increase the contrast for the moving cells. This could allow automated cell tracking even in the absence of fluorescent labels.
- PDMS devices are permeable to air and water, limiting the duration of experiments in the absence of an environmental chamber with humidity control.
- It is possible to prime the cell loading chamber directly with the cell suspension and use the air trapped in the migration channels to temporarily confine the cells to the loading chamber. The air could be subsequently be removed by increasing the fluid pressure in the device, after the cells have attached (Mills, Frith, Hudson, & Cooper-White, 2011).
- Immunochemistry of cells in channels is possible. One important difference compared to regular protocols on flat slides is the timing of each step.

Considering that the reagents will reach and leave the cells only by diffusion, it is important to increase the time with the square of molecular weight of the reagent (from minutes to hours for the fixation step, from minutes to tens of hours for the antibody loading and removing).

References

Ambravaneswaran, V., Wong, I. Y., Aranyosi, A. J., Toner, M., & Irimia, D. (2010). Directional decisions during neutrophil chemotaxis inside bifurcating channels. *Integrative Biology (Cambridge), 2,* 639–647.

Balzer, E. M., Tong, Z., Paul, C. D., Hung, W. C., Stroka, K. M., Boggs, A. E., et al. (2012). Physical confinement alters tumor cell adhesion and migration phenotypes. *FASEB Journal, 26,* 4045–4056.

Butler, K. L., Ambravaneswaran, V., Agrawal, N., Bilodeau, M., Toner, M., Tompkins, R. G., et al. (2010). Burn injury reduces neutrophil directional migration speed in microfluidic devices. *PLoS One, 5,* e11921.

Desai, S. P., Bhatia, S. N., Toner, M., & Irimia, D. (2013). Mitochondrial localization and the persistent migration of epithelial cancer cells. *Biophysics Journal, 104,* 2077–2088.

Faure-Andre, G., Vargas, P., Yuseff, M. I., Heuze, M., Diaz, J., Lankar, D., et al. (2008). Regulation of dendritic cell migration by CD74, the MHC class II-associated invariant chain. *Science, 322,* 1705–1710.

Gallego-Perez, D., Higuita-Castro, N., Denning, L., DeJesus, J., Dahl, K., Sarkar, A., et al. (2012). Microfabricated mimics of in vivo structural cues for the study of guided tumor cell migration. *Lab on a Chip, 12,* 4424–4432.

Irimia, D., Charras, G., Agrawal, N., Mitchison, T., & Toner, M. (2007). Polar stimulation and constrained cell migration in microfluidic channels. *Lab on a Chip, 7,* 1783–1790.

Irimia, D., & Toner, M. (2009). Spontaneous migration of cancer cells under conditions of mechanical confinement. *Integrative Biology (Cambridge), 1,* 506–512.

Jacobelli, J., Friedman, R. S., Conti, M. A., Lennon-Dumenil, A. M., Piel, M., Sorensen, C. M., et al. (2010). Confinement-optimized three-dimensional T cell amoeboid motility is modulated via myosin IIA-regulated adhesions. *Nature Immunology, 11,* 953–961.

Jones, C. N., Dalli, J., Dimisko, L., Wong, E., Serhan, C. N., & Irimia, D. (2012). Microfluidic chambers for monitoring leukocyte trafficking and humanized nano-proresolving medicines interactions. *Proceedings of the National Academy of Sciences of the United States of America, 109,* 20560–20565.

Kraning-Rush, C. M., Carey, S. P., Lampi, M. C., & Reinhart-King, C. A. (2013). Microfabricated collagen tracks facilitate single cell metastatic invasion in 3D. *Integrative Biology (Cambridge), 5,* 606–616.

Lammermann, T., Bader, B. L., Monkley, S. J., Worbs, T., Wedlich-Soldner, R., Hirsch, K., et al. (2008). Rapid leukocyte migration by integrin-independent flowing and squeezing. *Nature, 453,* 51–55.

Malawista, S. E., & de Boisfleury Chevance, A. (1997). Random locomotion and chemotaxis of human blood polymorphonuclear leukocytes (PMN) in the presence of EDTA: PMN in close quarters require neither leukocyte integrins nor external divalent cations. *Proceedings of the National Academy of Sciences of the United States of America, 94,* 11577–11582.

McDonald, B., Pittman, K., Menezes, G. B., Hirota, S. A., Slaba, I., Waterhouse, C. C., et al. (2010). Intravascular danger signals guide neutrophils to sites of sterile inflammation. *Science, 330,* 362–366.

Mills, R. J., Frith, J. E., Hudson, J. E., & Cooper-White, J. J. (2011). Effect of geometric challenges on cell migration. *Tissue Engineering Part C Methods, 17,* 999–1010.

Renkawitz, J., Schumann, K., Weber, M., Lammermann, T., Pflicke, H., Piel, M., et al. (2009). Adaptive force transmission in amoeboid cell migration. *Nature Cell Biology, 11,* 1438–1443.

Rolli, C. G., Seufferlein, T., Kemkemer, R., & Spatz, J. P. (2010). Impact of tumor cell cytoskeleton organization on invasiveness and migration: A microchannel-based approach. *PLoS One, 5,* e8726.

Scherber, C., Aranyosi, A. J., Kulemann, B., Thayer, S. P., Toner, M., Iliopoulos, O., et al. (2012). Epithelial cell guidance by self-generated EGF gradients. *Integrative Biology (Cambridge), 4,* 259–269.

Smolen, G. A., Zhang, J., Zubrowski, M. J., Edelman, E. J., Luo, B., Yu, M., et al. (2010). A genome-wide RNAi screen identifies multiple RSK-dependent regulators of cell migration. *Genes and Development, 24,* 2654–2665.

Wolfer, A., Wittner, B. S., Irimia, D., Flavin, R. J., Lupien, M., Gunawardane, R. N., et al. (2010). MYC regulation of a "poor-prognosis" metastatic cancer cell state. *Proceedings of the National Academy of Sciences of the United States of America, 107,* 3698–3703.

Micropatterned Porous Membranes for Combinatorial Cell-Based Assays

11

Clément Vulin, Fanny Evenou, Jean Marc Di Meglio, and Pascal Hersen

Laboratoire Matière et Systèmes Complexes, UMR7057, CNRS & Université Paris Diderot, Paris, France

CHAPTER OUTLINE

Abstract

Here, we describe a protocol for producing micropatterned porous membranes which can be used for combinatorial cell-based assays. We use contact printing to pattern the surface of a porous filter membrane with a thin layer of polydimethylsiloxane (PDMS). This allows the porosity of the filter membrane to be altered at selected locations. Cells can be grown on one side of the filter membrane, while drugs and reagents can be deposited on the porous areas of the other side of the membrane. The reagents can diffuse through the pores of the membrane to the cells. The first part of the protocol describes how to design a stamp and use it to contact print PDMS. The second part describes how to create microprinted membranes for cell-based assays. The method is simple, highly customizable, can be performed at the bench, and can be used to perform combinatorial or time-dependent cell-based assays.

INTRODUCTION

The combination of microfluidics and fluorescence microscopy has been instrumental in recent developments in quantitative cell biology (Bennett & Hasty, 2009; Sia & Whitesides, 2003). In particular, several microfluidics systems which allow the chemical environment of eukaryotic cells to be altered while simultaneously measuring the expression of key reporter genes by fluorescence microscopy have been proposed (Jovic, Howell, & Takayama, 2009; Taylor et al., 2009). Yet, these systems are still complex to use. A number of commercial systems exist; however, they are often limited in terms of operability and, without solid expertize, cannot be used to perform combinatorial and/or time-dependent cell-based assays (Castel, Pitaval, Debily, & Gidrol, 2006; Evenou, Di Meglio, Ladoux, & Hersen, 2012; Wu et al., 2010). Here, we describe the setup of a simple device to perform semi high-throughput assays on a monolayer of adherent cells. Our method is inspired by the fabrication method of "paper-based microfluidics" (Carrilho, Martinez, & Whitesides, 2009; Nery & Kubota, 2013), which aims to create cheap and "ready to use" diagnostic systems with the use of paper. Paper-based microfluidics are made from a porous material (usually paper, but other porous materials can be processed) on which a pattern is printed using a hydrophobic wax (Carrilho et al., 2009). This pattern defines hydrophobic and hydrophilic areas. Liquids can permeate the hydrophilic porous areas but cannot infiltrate where the wax has been printed. This approach is simple and does not require expensive equipment or materials. However, the delivery of drugs or metabolites to a pool of cells in such systems is limited (Derda et al., 2011). In particular, paper-based systems do not enable the possibility of repeated toxicity testing or pulse/chase experiments in cells. Indeed, liquids move by capillarity inside the porous meshwork and cannot be washed off or removed from the paper. Hence, such systems are usually designed for single use. Here, we use contact printing to deposit a thin layer of a biocompatible hydrophobic polymer (PDMS, polydimethylsiloxane) on top of a commercial filter membrane which has the properties of limited retention of liquids and proteins. In doing so, we modify the local porosity of the filter membrane (Evenou et al., 2012). Adherent cells can be cultured on the surface of the filter membrane, and chemicals can be delivered manually using a pipette from the other side of the membrane at selected locations where the porosity has not been blocked by the PDMS polymer. The resulting device, which we call a micropatterned membrane (μPM) is easy to produce and allows the effects of several drugs to be simultaneously tested on a monolayer of cells. We first describe the preparation of the porous substrate and then illustrate how it can be used for semi-high-throughput cell-based assays.

11.1 PATTERN DESIGN

The first step is to print the surface of a porous filter membrane with a thin layer of PDMS. This is performed using contact printing and requires the production of a master stamp; we have successfully used both direct micromachining and soft lithography (Whitesides, Ostuni, Takayama, Jiang, & Ingber, 2001; Xia & Whitesides, 1998) to fabricate stamps.

11.1.1 Pattern sizes and printing resolution

The design of the patterns will largely depend on the aim of the study. We have successfully printed different shapes—resulting in several patterns of porosity—ranging from simple lines and arrays of disks to more complex patterns. A few simple rules can help when drawing the patterns. First, the stamp should be made using PDMS, so that the uncured, liquid PDMS mixture to be used for contact printing will homogeneously spread on the surface of the stamp (i.e., good wettability). Once printed on the filter membrane, the thin layer of liquid PDMS will spread a little and blur the features of the pattern. This prevents the successful printing of fine details; sharp angles and tiny structures will not be faithfully reproduced. The printing protocol described later is limited to a typical resolution of 100 μm. A finer resolution can be obtained, but will require increased precision during the manipulation and deposition of the PDMS thin film. The nature of the filter membrane (wettability) and its porosity will also influence the final resolution of the pattern. The typical device that we use for cell-based assays consists of an 8×8 array of porous disks (1 mm in diameter), each separated by 1 mm nonporous areas. This pattern can be viewed as a miniaturized well-less plate, on top of which 2 μl drops of various reagents can be deposited to perform screening or cell-based assays on cells growing on the other side of the porous areas. Note that chemicals will diffuse within the media surrounding the cells, and therefore may act on cells further away than the cells immediately below the porous area on which the chemical has been deposited. This places constraints on the minimum distance between two porous areas and on the duration of the exposure time. In most cases, a distance of 1 mm between two porous areas is sufficient to avoid cross-talk between two consecutive porous areas; however, this distance may need to be adapted depending on the experimental protocol.

11.1.2 Materials

Several software programs can be used to draw a simple, geometrical pattern and export it in a format which can be used directly by a high resolution printer. Professional software programs (*AutoCAD* from Autodesk Inc., *L-Edit* from Tanner EDA) include advanced drawing features but are expensive and often not very user-friendly. A variety of open source software are also available and, in our opinion, these programs are ideal to start with. In particular, *Layout Editor software* is a powerful, free solution. A list of other layout editors can be found on their website (http://www.layouteditor.net/links/).

Printing companies now offer printing on transparent plastic sheets up to a resolution of 50,800 dpi, which allows the creation of patterns larger than 10 μm when using soft lithography. These companies usually work with the file format GDSII: a standard binary file format which is used to represent complex layouts. However, since the resolution of the final pattern on the membrane is limited by the contact printing—and not by the fabrication of the stamp—we do not need to achieve such high resolutions. It is actually possible to speed up the printing process and reduce costs by using a simple, standard inkjet printer. In this case, any bitmap or vector image editor can be used to draw the pattern and print it. We commonly use the commercial software Adobe Illustrator (Adobe Systems Incorporated) or Inkscape (open source GNU software) to draw patterns. We then use a Canon Pixma IP400 inkjet printer (Canon Ltd) at a resolution of 600 dpi and clear inkjet films (e.g., 100–075 from Mega Electronics) to produce the final photomask.

11.2 MASTER STAMP FABRICATION

The first step is to transfer the 2D printed pattern (the photomask) into a 3D pattern (the master stamp). Classically, this is performed using soft lithography. This is the limiting step in terms of resources and equipment, since you will need access to a clean room, or at the very least to a spin coater, a strong ultraviolet (UV) source, and heating plates. However, this equipment is only required to create the master stamp, which can be easily reused several times and replicated using epoxy resists or PDMS.

11.2.1 **Materials required for fabrication**

- Transparency photomask (see above, minimum size should cover the size of a porous filter membrane)
- SU-8-2100 photoresist and SU-8 developer (Microchem Corp)
- Isopropanol (for the developing steps)
- Silicon wafer (e.g., from SI-Mat)
- A metallic spatula for dispensing resist
- Two glass crystallizing dishes (typically 140–190 mm in diameter) for the rinsing steps.

11.2.2 **Equipment for fabrication (usually done in a clean room)**

- Spin coater (e.g., Delta +6RC from SUSS MicroTec, Germany) with a small vacuum pump to hold the sample (e.g., #FB65455 from ILMVAC, Germany)
- Two heating plates set at 65 and 95 °C
- UV lamp for resist exposure (e.g., Spectra Physics/Newport ref 66902 or MJB4 from SUSS MicroTec)
- A plasma cleaner (e.g., Diener Electronics)
- An air gun for drying the final wafer.

11.2.3 **Method**

Soft lithography has been described in several key articles (Whitesides et al., 2001; Xia & Whitesides, 1998). Soft lithography can be used to convert a 2D pattern into a 3D structure. When using SU-8 as a photoresist, the photomask should be opaque at the locations where PDMS will be contact printed onto the membrane (see Fig. 11.1). SU-8 is a negative photoresist, meaning that all regions of the photoresist layer that are illuminated by UV light will appear in relief after curing and development. We mainly use SU-8-2100 photoresist and follow the manufacturer's protocol (which takes ∼3 h) which can be found on their website (http://micro chem.com/pdf/SU-82000DataSheet2100and2150Ver5.pdf). This photoresist allows the creation of deep reliefs (>100 µm), which are suitable for contact printing of large areas. For our purposes, it is not advisable to make the patterns less than 100 µm thick, and we usually work with a 200–250 µm resist thickness. We quickly describe the manufacturer's protocol, focusing on several tips that can help when dealing with SU-8-2100, which is a very viscous resist.

- (Facultative) wafer preparation (step duration 10 min, total duration $t = 10$ min)
  - Clean the surface of the wafer using O_2 plasma at 6 mbar for 6 min.
  - Prior to deposition of the photoresist, the wafer can briefly be heated (95 °C) to encourage the spread of the photoresist.
- Resist spreading (step duration 10 min, $t = 20$ min)
  - Photoresists are usually stored at 4 °C. It is important to make sure that the resist is preheated to room temperature to achieve the nominal viscosity.

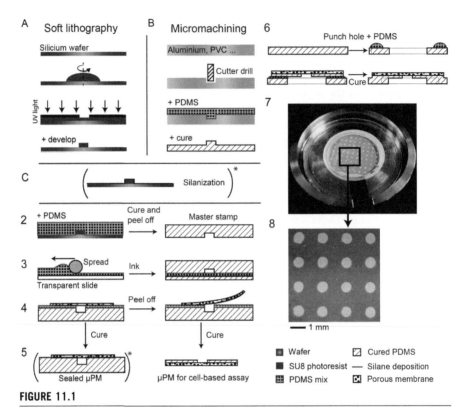

FIGURE 11.1

Patterned membrane fabrication. 1. Fabrication of the master pattern via soft lithography (A) or direct micromachining (B). (A) SU8 photoresist is evenly deposited on a flat silicon wafer by spin coating. The desired pattern is produced by UV exposure through a photomask followed by curing and development steps. (B) For simple patterns, one can also use micromachining and a hard material (such as metal or plastic). This can be performed using a microdriller or a microcutter drill. If required, the shape can be inverted by replica molding of PDMS. (C) A silanization step can be performed prior to master stamp molding; silanization is mandatory if the master pattern is created in PDMS (B), and advised if the substrate was produced by soft lithography to extend its lifetime. 2. The master stamp is fabricated by replica molding a 2–5 mm thick layer of PDMS on the master pattern. 3. A thin layer of uncured PDMS is spread on a flat surface such as printing transparency sheet. A tubular object such as a 10-ml pipette can be used to ensure homogeneous spreading. Ink the PDMS stamp by gently tapping it on the surface of the PDMS layer. 4. Turn the stamp upside down and place a porous filter membrane on top. Gently peel off the membrane and cure it to obtain a micropatterned membrane (μPM). 5. Direct curing can be used to obtain a microfluidic device with an integrated porous membrane, or the μPM can be cured alone on a hotplate at 100 °C. 6. To improve the ease of handling, a PDMS ring can be created by punching a hole in a 0.5–1 mm PDMS block. This ring is then attached to the μPM using uncured PDMS as a mortar followed by further curing. 7. A (metallic) holder can be used to facilitate cell culture. 8. View of the typical patterning of a μPM containing an array of millimeter scale porous discs. (See color plate.)

- Place the wafer on the spin coater and dispense around 5 ml of the photoresist on the wafer. If needed, use a spatula to spread the resist. SU-8-2100 is very viscous and it is important to initially help the spreading.
- Wait for the resist to spread on its own for a few minutes (typically 5 min).
- Spin the wafer in two steps (500 rpm for 120 s, 1500 rpm for 60 s). These steps need to be adjusted depending on the desired thickness of photoresist. The faster the rotation speed, the thinner the final photoresist layer. Note that the precise value of the thickness is not relevant here.
- Preexposure bake (step duration 46 min, $t = 1$ h 06 min)
  - Bake on hotplate for 6 min at 65 °C and then for 40 min at 95 °C. Do not allow the wafer to cool down between the two steps, as the gradual temperature increase is used to diminish mechanical tensions in the resist.
- Exposure (step duration 2 min, $t = 1$ h 08 min)
  - Place the transparency mask in contact with the photoresist layer after it has cooled down (this allows the photoresist to stiffen and become less sticky, thus preserving the mask). Expose to UV light (the exposure time depends on lamp power and focalization, and on the thickness of the pattern—we use 40 s at 10 mW/cm^2). Take into account that this time will also depend on the degree of UV absorption by the mask; a typical plastic mask will block 25% of the UV light. A good exposure time will result in a 90° edge on the photoresist/wafer interface, which is important for membrane patterning.
- Postexposure bake (step duration 18 min, $t = 1$ h 26 min)
  - Bake on hotplate for 5 min at 65 °C and then for 13 min at 95 °C. The resist should not wrinkle nor crack when heated. If this occurs, an intermediate 1 min temperature step (80 °C) can be added. In general, wrinkles should not be a problem for patterning, since such deformations will not affect the surface of the stamp that will be used for contact printing.
- Development and rising (18 min, $t = 1$ h 44 min)
  - Place the developer solution into a crystallizing dish and add the wafer. Agitate slowly for 16 min and then rinse the wafer with isopropanol in a new crystallizing dish. Isopropanol is best dispensed using a wash bottle. Carefully air-dry the wafer using compressed air (air gun). If white residue appears during the rinsing step, this means that either the development time was too short or that the developer bath is saturated with photoresist. Discard the entire contents of the developer bath in the proper manner and develop the wafer further in fresh developer solution.

11.2.4 **Alternative method: micromachining**

For simple patterns such as a single circular porous area or a straight line, the fabrication can be performed directly by micromachining of plastic or metallic materials. We routinely use polymethyl methacrylate (PMMA) or metallic substrates. Note that the features of the pattern must be sharply contoured in order to enable good pattern resolution. When drilling, do not go too deep (<1.5 mm), as it could then be difficult to replica mold the pattern using PDMS and/or use the PDMS replica for stamping due to too high an aspect ratio. Flat-bottomed millers are also preferred to classical drill bits. Note that replica molding the PMMA/metal piece with PDMS (Fig. 11.1) is advised. Indeed, the PMMA/metal piece can be used directly for stamping the membrane; however, this is not advisable as it usually yields poor printing quality due to the suboptimal wettability of the liquid to be printed (uncured PDMS). Also, it is often easier to use a flexible stamp rather than a rigid one (see the succeeding text). Modern fabrication techniques such as 3D printing could also be used to achieve more complex patterns without having to deal with soft lithography.

11.3 CREATING THE FINAL PDMS STAMP

11.3.1 Materials

- Silane (e.g., trichloro(1H,1H,2H,2H-perfluorooctyl) silane, #448931-10G from Sigma-Aldrich)—note that silanes are usually a dangerous product and should be handled with care, while wearing protective gloves, eye protection, and a lab coat).
- Pressurized argon gas (to keep silane under a neutral atmosphere).
- Small beaker or plastic cup.
- PDMS kit (Silgard184 from Dow Corning).
- Transfer pipette for dispensing silane.
- A glass dish that can hold the silicon wafer and resist a temperature of 65 °C.
- Scalpel.

11.3.2 Equipment

- Chemical fume hood and safety equipment for handling silane.
- Vacuum chamber (e.g., #10528861 from Fisher Scientific, France) and pump (e.g., Alcatel Vacuum Pump Type 2002BB).
- Precision scale (mg).
- Vacuum chamber and pump for degassing PDMS (as above, but note that a smaller pump can be used, e.g., #10661633 from Fisher Scientific, France).
- 65 °C oven.

11.3.3 Method

11.3.3.1 Silanization of the wafer (optional, in a chemical hood, use protective equipment at all times)

Silanization is required to ensure that the PDMS can be easily peeled off the silicon wafer without damaging it. In the absence of silanization, the photoresist layout on the wafer may break after molding a few PDMS replicas. Silanization is also required if using a master mold in PDMS (see Fig. 11.1B); however, it is not required when replica molding a PMMA/metal master directly.

- Place the wafer face up in a vacuum chamber. Beside the wafer, dispense three drops of silane in a small beaker (5 ml) or a plastic cup, and close the vacuum chamber.
- Apply a vacuum for 5 min, then stop the pump and let the wafer and silane sit under vacuum for 1 h to allow silane vaporization and deposition on the surface of the master pattern.
- Break the vacuum to recover the wafer. A good silanization process will leave little to no marks on the negative.

11.3.3.2 Casting PDMS on the wafer

- In a disposable plastic beaker (e.g., #11738549, Fisher Scientific, France) and using a precision scale, mix PDMS with its curing agent at a ratio of 10:1 by mass. Mix thoroughly using a spoon, a coffee stirrer stick, or a plastic pipette. The quantity of PDMS required will vary depending on the size of the stamp to be made. Easy-to-use thicknesses for the stamp are typically between 2 and 5 mm. A 3-in. wafer will fit into a 90-mm Petri dish and requires ~20–25 g of PDMS.
- Place the PDMS base and curing agent mixture in a vacuum chamber until all of the air bubbles are eliminated; the mixture will then be fully degassed. Note that the duration of this step may vary depending on the power of the vacuum pump. Also note that the PDMS liquid mixture will rise substantially due to bubble formation. To avoid overflowing, the volume of the beaker should be much larger than the volume of PDMS. We typically use a 150-ml beaker for 20–40 g of PDMS. Gently break the vacuum to recover the PDMS when no more bubbles are apparent.
- Carefully pour the PDMS over the stamp, face up. Incubate at 65 °C for at least 3 h; it is also possible to cure overnight.
- To peel off the stamp, carefully cut the PDMS around the zone of interest using a scalpel and gently pull the stamp away. Take care not to cut too close to the patterns, as the membrane surface must fit within the surface of the stamp, and be gentle so as not to break the silicon wafer.

11.4 PATTERNED MEMBRANE FABRICATION
11.4.1 Membrane selection

Membrane patterning has been successfully applied to several commercially available membranes or filtration papers. Among those that we often use are polycarbonate track etched membranes (e.g., Isopore™ #GTTP02500—25 mm diameter; Millipore) which are the preferred choice in our lab; alumina-based filtration membranes (e.g., Anodisc™ #6809-6022; Whatman) which are very porous, but brittle; or polyester-based membranes (Cyclopore™, #7060-2502; Whatman). These filtration membranes are available in different sizes (we usually use 25 mm in diameter) and with different pore sizes. Larger pores will facilitate increased diffusion of the reagents. Also, note that these membranes are opaque; therefore, the cells cultured on them can only be observed using epifluorescent microscopy. Transparent (less porous) membranes (e.g., Cyclopore™ from Whatman) also exist and may facilitate standard microscopy observations. There are three main criteria to consider when choosing a membrane:

- First, the membrane should be compatible with cell culture. Due to the size of pores, membrane ruggedness, or other factors, cells may not attach or may have difficulty growing on the membrane surface. To promote cell culture, the membrane surface should be coated with collagen, fibronectin, or other proteins that facilitate cell adhesion; alternatively one can use a thin layer of Matrigel™ (from BD Biosciences).
- Second, the membrane should be easy to pattern. Thin (<100–200 μm), chemically and mechanically resistant and highly porous membranes are the optimal choice, provided PDMS wettability is good. A high absorption capacity (e.g., Anodisc™) leads to a higher printing resolution, but may result in thickness printing heterogeneities in certain zones. Also, the membranes should be resistant to high temperatures (>100 °C) to enable rapid curing of the PDMS.
- Lastly, membrane porosity should be selected on the basis of the size and affinity of the diffusing drug molecules, and with respect to the possibility of cell protrusion through the membrane. We primarily use 0.2 and 1 μm pore sizes. A membrane with cross-linked pores or pores which are nonperpendicular to the membrane may be used, but will lead to a lower spatial precision in cell-based assays.

11.4.2 Materials for membrane patterning

- Filtration membranes (see Section 11.4.1 for references).
- Flat-tipped tweezers for membrane manipulation.
- Reusable spacers to avoid membrane/hot plate contact. We typically use 20 mm vulcanized fiber plumbing gaskets as a cheap source of spacer; these can be found in most hardware shops.
- Two to three grams of premixed/degassed PDMS (see Section 11.3.3.2). Sylgard 184 PDMS (Dow Corning) is a natural choice; however, it is often appropriate to use black PDMS (Sylgard 170, Dow Corning).
- A 10 cm × 20 cm piece of printing transparency sheet, or equivalently flat, smooth plastic for homogeneous spreading of the PDMS.
- A 10 cm smooth rod to spread PDMS (e.g., a 10 ml plastic pipette tip or a glass rod).
- Aluminum foil or an appropriate crystallizing dish to cover and protect the membranes during curing.
- A hotplate set at the appropriate temperature (e.g., 100 °C).

11.4.3 **Method**

- Pour a few drops of PDMS on a sheet of printing transparency. Spread the PDMS using a rod to obtain an even layer approximately one-tenth of a millimeter thick. For this step, the use of black PDMS is ideal since it allows direct visualization of the homogeneity of spreading.
- Ink the stamp by placing the stamp upside down on the PDMS layer and gently tapping the stamp. This step is crucial to achieve good patterning. The stamp should be coated in an even, thin layer of PDMS, with an absence of edge beads or PDMS inside patterns. If the inking leads to an unequal layer of PDMS on the stamp, it is advised to wait a few seconds to allow the layer to equilibrate. In practice, it may take a few spreading/inking attempts to obtain an appropriate layer of PDMS. In this case, the stamp surface can simply be wiped clean using tissue paper, and if required cleaned with acetone and carefully dried before the process can be started again.
- Carefully place the membrane on top of the stamp. Wait a few seconds for total contact to occur and peel off the membrane using a pair of tweezers. Avoid sliding the membrane when peeling it off; this is highly likely to destroy the printed PDMS layer. When using brittle membranes such as Anodiscs™, the flexibility of the stamp is crucial to help membrane separation: in this case, the stamp is peeled off while the membrane remains untouched.
- Note that it is also possible to leave the membrane on top of the stamp and to cure it directly. We will not provide further details here; however, this is a simple method of including a porous filter in a microfluidics system.
- Quickly place the membrane inked side up on a spacer sitting on the hotplate (the inked side will hereafter be referred to as the recto side). The optimal temperature is around 100–120 °C, as it allows rapid curing of the PDMS and ensures a good pattern resolution. Most porous filtration membranes can resist such temperatures; however, some may be altered. In this case, the temperature can be lowered.
- Allow curing for 1 h. To avoid dust deposition, the membranes should be covered using aluminum foil or a crystallizing dish.
- The membranes can then be stored in a clean/dust free box.

11.4.4 **Tip for membrane handling**

To ease handling, the membranes can be attached to a 1-mm-thick PDMS washer. To do so, a 1 mm layer of PDMS mixture (see Fig. 11.1) can be poured in a Petri dish, cured at 65 °C overnight and then cut to the appropriate size using a scalpel. The PDMS washer can then be glued onto the inked side of the patterned membrane using drops of uncured PDMS as a mortar, followed by further curing at 65 °C.

11.5 MEMBRANE PREPARATION FOR CELL-BASED ASSAYS

11.5.1 Materials and equipment required for membrane preparation

- Round-tipped tweezers for membrane manipulation.
- UV lamp for sterilization treatment.
- Twenty milliliters of phosphate buffered saline (PBS).
- Five milliliters of coating solution to facilitate cell attachment to the porous filter, for example, 0.3 g/l type I collagen, fibronectin, or the relevant adhesion ligand for your chosen cell line.
- Five milliliters of sterile deionized water.

11.5.2 Method

- Membranes should be washed with PBS to remove any dust and should then be sterilized on both sides using UV light. This can be accomplished using a dedicated UV source. We usually expose the membranes for 30 min on each side to the UV light of a conventional cell culture hood.
- Incubate with the desired coating protein—for example, 0.3 g/l type I collagen—for 2 h at room temperature.
- Rinse once with water and twice with PBS.

11.5.3 Cell seeding

Prior to cell seeding, we routinely culture MadinDarby Canine Kidney (MDCK) cells in 25 cm^2 tissue culture flasks (Techno Plastic Products AG (TPP), Switzerland) in a 37 °C incubator with a 5% CO_2 atmosphere. The culture medium is Dulbecco's Modified Eagle Medium supplemented with 10% fetal bovine serum, 100 units/ml penicillin and streptomycin (Gibco), and 100 μg/ml kanamycin (Sigma-Aldrich). The MDCK cells are trypsinized, transferred for culture on the collagen-coated verso side of the μPMs, and allowed to grow to near confluence (2 or 3 days) in a 37 °C incubator with a 5% CO_2 atmosphere before the assay starts. Note that the best results are obtained when the cells are cultured on the verso side, possibly due to surface irregularities and the hydrophobicity of PDMS. It is also better to deposit the drops of the reagents to be tested on the printed PDMS side, as the hydrophobicity of PDMS ensures the drops of reagent are tightly confined which prevents cross-contamination. Most cell lines will be suitable for culture on μPMs, as long as they can adhere to the coated surface. We have tested MDCK, HeLa, and 3T3 cell lines. Again, note that μPMs are not transparent; therefore, the cells can only be observed by fluorescence microscopy. This requires the use of either cell lines transfected with a fluorescent reporter gene or staining with fluorescent dyes appropriate to the cellular process under study.

11.5.4 Method

- Take the μPM and rinse the cells with fresh culture medium.
- To avoid direct contact between the cells and the Petri dish, it is advisable to use a holder to support the μPM (Fig. 11.1). It can easily be made in PDMS by

punching a hole into a 1–2 mm thick PDMS layer. If possible it is more convenient to make a holder in stainless steel that can be autoclaved and reused.

- Mount the μPM on the holder, with the recto side up so that the cells are immersed in the culture medium, with the top of the membrane at the surface; we use 4 ml of culture medium in 35-mm dishes. To facilitate observation of the cells, one can use 35-mm Petri dishes with glass bottom (e.g., #81158 from iBidi).
- Dispense 1–2 μl drops (for 1 mm porous disk) of the reagents to be tested at separate locations on the recto side of the μPM.
- Let the cells incubate in a 37 °C incubator with a 5% CO_2 atmosphere for the desired period of time.
- Using a pipette, replace the drops of reagents with a drop of culture medium to stop the assay.
- If required, repeat the deposition of the drops of reagents to be tested for a time-course experiment or a combinatorial assay. In the case of repeated assays at the same locations, the first drops can be aspirated using a micropipette, and the porous disks rinsed twice with drops of culture medium before depositing the next drops of reagents.
- Image the cells using fluorescence microscopy.

Note that the cells will remain upside down, and that imaging of the cells during this process will need to be done using an inverted microscope. Since the cells are not directly in contact with the bottom of the Petri dish, it is not possible to use a high magnification, and objectives with a long working distance are preferred. The drops of reagents can be manually added or removed at any time during image acquisition. Thus, it is possible to perform time-lapse microscopy. However, the user needs to be careful to not move the sample or the microscope stage when placing the drops. As a proof of principle, we were able to manually stain cells from the same monolayer at specific locations with different dyes using this method (see Fig. 11.2). It is also possible to stack different membranes in order to change the pattern of porosity seen by the cells as illustrated in Fig. 11.2. That way, complex drug delivery, with variations in space and time can be tested manually.

CONCLUSION

We have described a simple method to contact print a thin layer of PDMS onto a porous substrate. This method can be used to create arrays of millimeter scale porous areas which can be used as a substrate for cell-based assays. Compared to 96-well plates and other cell-based assay strategies, our method has several advantages: it is highly customizable, inexpensive, easy to fabricate and manipulate, and allows complex combinatorial and time-dependent assays to be performed on a monolayer of cells. The same technique can be applied to systems other than mammalian cells, for example, microorganisms for antibiotic or metabolic screening.

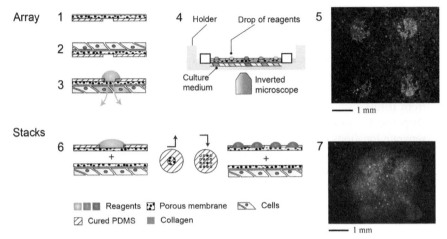

FIGURE 11.2

Cell culture assay. 1. Micropatterned membranes are coated with collagen prior to cell seeding. 2. Cells are grown to mid-confluence on the coated side of the membrane under normal cell culture conditions. 3. The μPM is then flipped upside down to allow the user to deposit drops of reagents on the porous areas. The drugs will selectively diffuse through the porous areas of the μPM to the cells growing immediately below. 4, 5. Multiple reagents can be tested at once, and their effects observed using an inverted fluorescent microscope. The hydrophobicity of PDMS allows precise deposition of the drops of reagents and selective cell stimulation. 6, 7. Stacking a nonpatterned membrane followed by a patterned membrane will allow the testing of multiple drugs or chemicals in different areas, in a combinatorial manner, by changing the μPM. (See color plate.)

References

Bennett, M. R., & Hasty, J. (2009). Microfluidic devices for measuring gene network dynamics in single cells. *Nature Reviews Genetics, 10*, 628–638.

Carrilho, E., Martinez, A. W., & Whitesides, G. M. (2009). Understanding wax printing: A simple micropatterning process for paper-based microfluidics. *Analytical Chemistry, 81*, 7091–7095.

Castel, D., Pitaval, A., Debily, M.-A., & Gidrol, X. (2006). Cell microarrays in drug discovery. *Drug Discovery Today, 11*, 616–622.

Derda, R., Tang, S. K. Y., Laromaine, A., Mosadegh, B., Hong, E., Mwangi, M., et al. (2011). Multizone paper platform for 3D cell cultures. *PLoS One, 6*, e18940.

Evenou, F., Di Meglio, J.-M., Ladoux, B., & Hersen, P. (2012). Micro-patterned porous substrates for cell-based assays. *Lab on a Chip, 12*, 1717–1722. http://dx.doi.org/10.1039/c2lc20696j.

Jovic, A., Howell, B., & Takayama, S. (2009). Timing is everything: Using fluidics to understand the role of temporal dynamics in cellular systems. *Microfluid Nanofluid, 6*, 717–729.

Nery, E. W., & Kubota, L. T. (2013). Sensing approaches on paper-based devices: a review. *Analytical and bioanalytical chemistry*, *405*, 7573–7595. http://dx.doi.org/10.1007/s00216-013-6911-4.

Sia, S. K., & Whitesides, G. M. (2003). Microfluidic devices fabricated in poly(dimethylsiloxane) for biological studies. *Electrophoresis*, *24*, 3563–3576.

Taylor, R. J., Falconnet, D., Niemisto, A., Ramsey, S. A., Prinz, S., Shmulevich, I., et al. (2009). Dynamic analysis of MAPK signaling using a high-throughput microfluidic single-cell imaging platform. *Proceedings of the National Academy of Sciences of the United States of America*, *106*, 3758–3763.

Whitesides, G., Ostuni, E., Takayama, S., Jiang, X., & Ingber, D. E. (2001). Soft lithography in biology and biochemistry. *Annual Review of Biomedical Engineering*, *3*, 335–373.

Wu, J., Wheeldon, I., Guo, Y., Lu, T., Du, Y., Wang, B., et al. (2010). A sandwiched microarray platform for benchtop cell-based high throughput screening. *Biomaterials*, *32*, 841–848.

Xia, Y., & Whitesides, G. M. (1998). Soft lithography. *Annual Review of Materials Science*, *28*, 153–184.

Micropatterning Cells on Permeable Membrane Filters

12

Sahar Javaherian*, Ana C. Paz*,†, and Alison P. McGuigan*,†

*Department of Chemical Engineering and Applied Chemistry, University of Toronto, Toronto, ON, Canada

†Institute for Biomaterials and Biomedical Engineering, University of Toronto, Toronto, ON, Canada

CHAPTER OUTLINE

ISSN 0091-679X
http://dx.doi.org/10.1016/B978-0-12-800281-0.00012-9

Abstract

Epithelium is abundantly present in the human body as it lines most major organs. Therefore, ensuring the proper function of epithelium is pivotal for successfully engineering whole organ replacements. An important characteristic of mature epithelium is apical–basal polarization which can be obtained using the air–liquid interface (ALI) culture system. Micropatterning is a widely used bioengineering strategy to spatially control the location and organization of cells on tissue culture substrates. Micropatterning is therefore an interesting method for generating patterned epithelium. Enabling micropatterning of epithelial cells however requires micropatterning methods that are designed to (i) be compatible with permeable membranes substrates and (ii) allow prolonged culture of patterned cells, both of which are required for appropriate epithelial apical–basal polarization. Here, we describe a number of methods we have developed for generating monoculture as well as coculture of epithelial cells that are compatible with ALI culture.

INTRODUCTION

Epithelium is one of the four major tissue components of the human body and lines the cavities and surfaces of most of our major organs (e.g., intestine, lung, kidney). Incorporating functional and mature epithelium into the design of engineered tissue is therefore key for generating tissue replacements with correct architecture and function (Soleas, Paz, Marcus, McGuigan, & Waddell, 2012). Additionally, more than 90% of human cancers arise in epithelial tissues (Frank, 2007). Accordingly, having more realistic *in vitro* models of epithelium, which reflect the complexity of the native tissue organization, could provide a better tool for understanding the pathology of epithelial diseases as well as provide a better culture platform for drug discovery aimed at battling these diseases.

Epithelial cells represent an example of polarized cells. Specifically epithelial cells can be polarized either in the apical–basal (Gibson & Perrimon, 2003; St Johnston & Sanson, 2011) or planar (Fanto & McNeill, 2004) axis of the cell. While only some epithelial cells exhibit planar polarization, all epithelial cells exhibit apical–basal polarization (Tepass, 2012). Apical–basal polarization involves the formation of apical and basal membrane domains with distinct molecular composition and function. The apical membrane domain is enriched in Bazooka (Par3)/Par6/aPKC and Crumbs (CRB1)/Stardust (Pals1)/PATJ complexes (Tepass & Knust, 1993). The apical domain acquires specific functional features critical for epithelium function, for example the formation of specialized structures such as primary and motile cilia and microvilli (Soleas et al., 2012). The basolateral membrane domain is located furthest from the lumen and is enriched in Lgl, DLg, and Scrib proteins as well as integrins (Kaplan, Liu, & Tolwinski, 2009). This domain facilitates adhesion of the epithelium to the underlying basal membrane.

In vitro, apical–basal polarization of epithelial cells is promoted using air–liquid interface (ALI) culture. In this culture system, cells are seeded on a permeable support (e.g., Transwell™) that separates the bottom and top chambers of the culture well. The permeable support is designed to allow diffusion of nutrients from the basal side. After a few days of culture in this system with medium in both the top and bottom chambers to allow the generation of a confluent cell sheet on the membrane, the medium is removed from the top chamber of the system. As a result, the basal membrane of the cells is exposed to medium and nutrients and the apical membrane is exposed to air. This culture system induces specialization of basolateral and apical membrane domains of the epithelial cells, ultimately leading to apical–basal polarization and maturation of the *in vitro* epithelium. Therefore, preserving this permeable support is important when designing cell-patterning approaches if they are to be fully exploited to pattern epithelial cells. Micropatterning is a widely used bioengineering strategy to spatially control the location and organization of cells on tissue culture substrates (Folch et al., 1999; Gomez et al., 2010; Janvier et al., 1997; Javaherian et al., 2011; Lang et al., 2006). Recently, our lab has developed and adapted existing micropatterning approaches originally designed for use on solid substrates for use with epithelial cells undergoing ALI culture (Paz, Javaherian, & McGuigan, 2011, 2012). We will present the detailed protocols for these methods (overview of each method is shown in Fig. 12.1) in this chapter with the hope of advancing the use of micropatterning in epithelial biology labs with minimal micropatterning experience.

We will present three methods, each with its own advantages, limitations, and range of applications. The first method, agarose microwells, is designed to generate micropatterned colonies of desired shape and size. The second method, extracellular matrix (ECM) microcontact printing, allows generation of patterned cocultures of two cell types. Finally, the third method, parafilm patterning, allows simple and fast generation of monocultures as well as cocultures of limited size and geometries.

Each method starts with prepatterning of the permeable support culture surface for spatial control of the cells. In the case of cocultures, once the first cell type is patterned on a prepatterned surface the cell-free areas of the permeable support are rendered cell adhesive and second cell type is seeded on them. In all cases, the timing of each step and the seeding densities required to achieve robust patterning require optimization for the specific cell type being used. The specifics provided here are for the cell types we have patterned but we recommend users optimize these parameters for their specific cell type of interest.

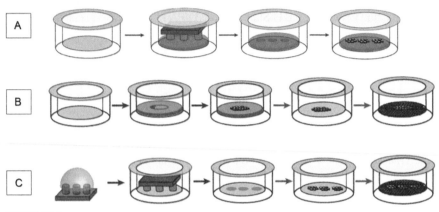

FIGURE 12.1

Schematic of patterning methods. (A) Agarose microwells method. A stamp prepared using soft lithography is placed on a permeable membrane. A solution of agarose is applied around the stamp. When agarose has gelled, the stamp is removed resulting in microwells in agarose. When cell suspension is applied to the insert, cells can only adhere in agarose-free microwells, resulting in patterned colonies. (B) Parafilm patterning method. A parafilm insert containing holes of desired shapes is bonded to the permeable membrane. When cell suspension is applied to the insert, cells can only adhere in parafilm-free areas, resulting in patterned colonies. Parafilm can be placed in place for monocultures or can be removed to generate patterned cocultures. For the latter, after removal of parafilm, the membrane surface can be reactivated using a solution of ECM to deem it cell adhesive. The seeding of second cell type (red) followed by removal of unadhered cells results in a patterned coculture. (C) ECM microcontact printing method. A stamp prepared using soft lithography is coated with a solution of ECM. Next, the ECM-coated stamp is placed on the permeable support and the ECM microprinted on the membranes surface. Incubation of the membrane in BSA solution ensures that only areas printed with ECM protein are cell adhesive. As a result, when cell suspension is applied to the insert, cells can only adhere in ECM printed areas, resulting in patterned colonies. Next, the membrane surface free of first cell type can be reactivated using a solution of ECM to deem it cell adhesive. The seeding of second cell type followed by removal of unadhered cells results in a patterned coculture. (See color plate.)

12.1 **MATERIALS**

12.1.1 **General to all three methods**

1. Permeable support culture system (e.g., Transwell^{TM} filters, BD Falcon, catalog # 29442-120)
2. Phosphate buffered saline (PBS)
3. Complete cell culture medium (specific to cell type being used)
4. Serum-free medium (specific to cell type being used)
5. UV sterilization cabinet (biosafety cabinet equipped with a UV light source can be used instead)
6. Tweezers/forceps

12.1.2 **Specific to agarose microwells and ECM microstamping methods**

1. Transparent sheet photomask with desired geometric features
2. Glass slide (VWR, catalog # 48312-024)
3. Plasma cleaner
4. Spin coater
5. Photoresist (e.g., SU-8, MicroChem, catalog # SU-8 25)
6. UV source
7. SU-8 Developer solution (MicroChem, catalog # SU-8 Developer solution)
8. Acetone (Sigma Aldrich, catalog # 320110)
9. Isopropanol (Sigma Aldrich, catalog # I9516)
10. 1,1,2,2-Tetrahydrooctyl-tridecafluorotrichlorosilane (United Chemical Technologies, catalog # T2492)
11. Vacuum chamber
12. Polydimethylsiloxane prepolymer and curing agent (10:1 mass-to-mass ratio) (Sylgard 184, Dow Corning, catalog # depends on local distributer)
13. Petri dish

12.1.3 **Specific to ECM microstamping and parafilm patterning methods**

1. 10 µg/ml stabilized bovine fibronectin (BTI Biomedical Technologies Inc., catalog # BT-226S) in complete growth medium

12.1.4 **Specific to agarose microwells method**

1. Low melting point agarose type II (ISC BioExpress, catalog # 0815-100G)
2. Anhydrous ethanol
3. 25 µg/ml stabilized bovine fibronectin solution

12.1.5 Specific to ECM microcontact printing method

1. 100 μg/ml stabilized bovine fibronectin in PBS
2. 0.05% (wt/v) solution of bovine serum albumin (Sigma Aldrich, catalog # A9418-50G) in ddH$_2$O
3. Glass cover slips (VWR, catalog # 48366-067-1)
4. KimWipes
5. 10-cm polystyrene dish
6. 6 g weights

12.1.6 Specific to parafilm patterning method

1. Parafilm^{TM}
2. Blunt end needles of desired diameter (e.g., Ameritronics, catalog # ZT-5-026-1-L)
3. Two scalpel blades (Magna Scalpel Blades, catalog # 2580-M90-10) taped together
4. Hot plate set at 50 °C

12.2 METHODS

12.2.1 Stamp and stencil fabrication

12.2.1.1 Stamp fabrication (specific to agarose microwells and ECM microcontact printing methods)

For the fabrication of the elastomeric stamp, first a SU-8 master containing specific topographic features has to be made. The master is made using conventional photolitography techniques (Nelson, Liu, & Chen, 2007; Whitesides, Ostuni, Takayama, Jiang, & Ingber, 2001).

  i. First, design a photomask with the desired geometric patterns using AutoCad, or any drafting software, and print the pattern in high-resolution black ink on a transparent sheet (any local printing company that can do high-resolution printing can be used to generate the mask).
 ii. Rinse a 70 × 50 mm glass slide with isopropanol, then with acetone, and finally with isopropanol again. Bake the slide at 100 °C overnight. This process is to dehydrate the slide.
iii. Clean the slide in the plasma cleaner for 1 min on the high-energy setting.
 iv. Spincoat over the clean glass slide a thin layer (15 μm thick) of negative photoresist US8-25 using a speed of 3000 rpm for 30 s. Place the glass slide on a hot plate (soft bake) at 65 °C for 3 min, 95 °C for 7 min, and 65 °C for 2 min. This step is to evaporate solvent and set the photoresist.
  v. Expose the glass slide to UV light using an exposure energy of 170 mJ/cm^2 to polymerize the photoresist. The time of exposure can be calculated using

the following formula: exposure energy/intensity = time (the intensity depends on the UV light source).

vi. Place the glass slide on a hot plate (postexposure bake) at 65 °C for 1 min, 95 °C for 3 min, and 65 °C for 2 min. This first photoresist layer is to promote the attachment of the second photoresist layer to the glass slide.

vii. Next spincoat over the glass slide a second layer of negative photoresist US8 (e.g., US8-2025) using a specific rate and duration, which depends on the desired height of the stamp posts. This parameter can be found on the photoresist data sheet. We used a speed of 1000 rpm for 30 s to get a thickness of 80 μm.

viii. Place the glass slide on a hot plate (soft bake). In this case the time and temperature depends on the desired photoresist thickness. Also, this parameter can be found on the photoresist data sheet. We applied 65 °C for 3 min, 95 °C for 8 min, and 65 °C for 2 min.

ix. Place the photomask over the cool glass slide and expose it to UV light using a specific exposure energy and time to selectively polymerize the exposed portion of the photoresist. The exposure energy depends on the desired height of the stamp posts and it is specified on the photoresist data sheet. We used an exposure energy of 97 mJ/cm^2.

x. Remove the photomask and place the glass slide on a hot plate (postexposure bake). We applied 65 °C for 2 min, 95 °C for 6 min, and 65 °C for 2 min. Again, this parameter can be found on the photoresist data sheet.

xi. Finally, wash away any unexposed regions of photoresist by submerging the glass slide in developer solution for 10 min. Wash the master with isopropanol and let it dry.

xii. To generate the elastomeric stamp silanize the master under vacuum for 3 h using 1-trichlorosilane. Place the silanized master with the feature side up on a 150 × 25 mm Petri dish covered with aluminum foil. Pour 50 g of 10:1 (w/w) mixture of Polydimethylsiloxane (PDMS) polymer:curing agent over master, degas the PDMS for 10 min in the vacuum, and cure at 60 °C for 3 h. Let the PDMS cool down and then peel off the PDMS from the master. Cut the PDMS stamp with a blade to get the desired stamp size.

12.2.1.2 Parafilm insert fabrication (specific to parafilm patterning methods)

i. Start by using a biopsy punch to cut a circular piece of parafilm slightly smaller than the diameter of the transwell to be used.

ii. Place the parafilm piece on a glass slide taking care not to stretch it.

iii. Using a blunt end needle make holes in the parafilm by bringing the needle down straight and applying gentle and uniform pressure (do not turn the needles as this will stretch parafilm). If rectangular shape cell patterns are desired, tape two scalpel blades together ensuring they are aligned properly. The distance between blades can be varied by placing a piece of laboratory tape or paper between blades before taping them together. Cut the parafilm using the generated double blade by firmly pressing down and dragging the blades across the surface over the length to be cut (if the blades are not sharp, this will stretch parafilm and compromise patterning). Cut the ends between two incisions in parafilm and remove the piece of parafilm using tweezers to leave behind a rectangular hole in parafilm.

12.2.2 Substrate prepatterning

12.2.2.1 Generating agarose microwells (specific to agarose microwells method)

    i. Place a filter well insert (e.g., Transwell™) into an empty well of a multiwell plate.

    ii. Using forceps, place a PDMS stamp on the filter membrane feature side down. Apply gentle pressure to ensure contact between the membrane and the post features of the PDMS stamp.

    iii. Prepare a 1% (wt/v) solution of agarose in PBS. Ensure that agarose is completely dissolved in the solution by bringing the solution to boiling.

    iv. While still hot, mix three parts of 1% agarose with two parts of anhydrous ethanol. Immediately perfuse the solution under the PDMS stamp. This step must be carried out while the solution is still hot to prevent premature evaporation of ethanol. The perfusion is easiest to carry out by placing a droplet of the solution on the filter well membrane at the corner of the stamp. The capillary forces will drag the solution under the stamp; if the stamp is not completely perfused, apply more agarose/ethanol solution to other corners of the stamp.

    v. Next, leaving the stamp in place, gel the agarose by incubating in the freezer ($-20\ ^{\circ}$C) for 3 min and then at room temperature for 5 min.

    vi. Finally, carefully remove the PDMS stamp with forceps by pulling straight up to generate the patterned agarose layer on the filter insert membrane. Note it is important to pull straight up to remove the stamp to prevent pattern distortion.

    vii. Sterilize the gel patterned inserts under UV for 30 min or more.

    viii. Coat the surface with 25 µg/ml fibronectin (use enough solution to completely cover the agarose microwells) for 1 h at room temperature. Remove fibronectin solution immediately before cell seeding without allowing the surface to dry.

12.2.2.2 ECM microstamping (specific to ECM microcontact printing method)

    i. Prepare a solution of the desired ECM stamping protein. Here, we use a solution of 100 µg/ml fibronectin, but other proteins have been also successfully used (e.g., collagen, gelatin).

    ii. Place each stamp on a glass cover slip feature side facing up. Cover slips are used here for making the handling of the stamps easier.

    iii. Cover the bottom of a 10-cm dish with wet KimWipes and place the cover slips with stamps into the dish.

    iv. Apply a solution of ECM protein on top of each slide making sure to cover the entire surface of the stamp (for a 5×5 mm stamp we use 50 µl of fibronectin solution). Place the lid on the dish to prevent evaporation and incubate at room temperature for 1 h.

v. After the incubation, using forceps lift the cover slip with a stamp on it and gently taking care not to touch the surface of the stamp wick the solution of ECM away using a KimWipe.

vi. Place a droplet of PBS large enough to cover the entire stamp surface on the stamp then wick the droplet away gently using a KimWipe. Ensure that the stamp is completely dry before proceeding. For certain pattern geometries, it might be beneficial to dry the stamp using a gentle stream of air.

vii. Remove the stamp from the cover slip using forceps and place it feature size down on a filter insert membrane. The filter insert membrane should be placed in an empty well of a multiwell plate for this step. It is important to carry this step out with great care. Once the stamp has touched the filter membrane, moving it will result in poor patterning. As such, we recommend dropping the stamp feature side down onto the filter membrane and not moving it further. Place a 6-g weight on top of the stamp and incubate for 10 min.

viii. Next, remove the stamp by lifting it straight up using forceps. Note the stamps are reusable and can be washed by sonication in water supplemented with detergent.

ix. Next place a solution of 0.05% bovine serum albumin (BSA) in the filter insert to cover the entire surface of the membrane. Incubate the prepatterned membrane in BSA solution for 1 h at room temperature to generate a cell-repellent surface in regions of the membrane not stamped with ECM protein. The BSA solution should be removed immediately before cell seeding without allowing the surface to dry up.

12.2.2.3 Parafilm annealing (specific to parafilm patterning method)

i. Place the prepared parafilm membrane into the filter well insert. Place the insert on a solid surface and press parafilm down firmly applying pressure to all parts of the membrane to ensure uniform annealing.

ii. Next, move the insert onto a hot plate set at 50 °C and heat while applying pressure for 5 s. Great care should be taken to prevent melting of the parafilm. Once the parafilm insert is annealed to the transwell membrane, place the transwell into a multiwall plate and fill the top chamber with PBS.

iii. Degas the membranes at 30 psig to ensure no air bubbles are trapped in the parafilm holes. This step can also be used to evaluate the quality of parafilm annealing to transwell membrane; if the parafilm is not properly bonded to the insert, it will come off during degassing. In case of that happening, a new parafilm insert needs to be fabricated.

iv. Following degassing, sterilize inserts under UV for 30 min or more.

v. Next, remove PBS from the inserts and incubate the prepatterned inserts with fetal bovine serum (FBS) to enhance cell adhesion. For certain cell types, FBS may need substituted for an ECM protein such as collagen. Immediately before cell seeding remove FBS without allowing the surface to dry up.

12.2.3 Seeding of the first cell type

For patterning monocultures of cells into colonies of desired shape with either agarose microwells or parafilm patterning methods this is the only cell seeding step. All the following steps should be carried out in a biosafety cabinet under sterile conditions.

i. Trypsinize cells and make sure no cell aggregates are present in the solution.

ii. Centrifuge the cell suspension and resuspend the cells in serum-free medium at a concentration of 2×10^6 cells/ml, making sure there are no cell aggregates in the resuspended solution.

iii. Apply the cell suspension onto the prepatterned filter well inserts. For a six-well insert we use a total of 1 ml per insert, resulting in 1×10^6 cells in a filter well insert. The success of cell patterning is very sensitive to this cell seeding number. If a filter well insert of a different size is used, the cell number should be readjusted based on the surface area of the membrane (keeping the cell seeding density at around 2200 cells/mm^2). Allow the cells to settle (around 3–5 min) and visually ensure that 100% of the membrane surface is covered by cells. Without disrupting the cells move the well insert into an incubator and incubate for 30 min (this incubation step can be increased to 1 h for monoculture patterning).

iv. Next remove unadhered cells and gently wash the patterns with serum-free medium 4–5 times. Generated patterns should be clearly visible at this stage. If patterning is successful, you should see confluent cell patterns (Fig. 12.2A).

v. In the case of monoculture patterning using agarose microwells or parafilm pattering method, place appropriate volumes of complete growth medium in both chambers of the filter insert system (volume should be matched, following filter insert manufacturer's guidelines, to prevent excessive pressure across the membrane and detachment of the cells) and continue culture until the desired time to start ALI culture (Section 12.2.6). A representative example of monoculture pattern obtainable by agarose microwells method is shown in Fig. 12.3.

12.2.4 Converting cell-free areas of filter well insert to become cell adhesive

This step is to be carried out only in case of patterning cocultures.

12.2.4.1 ECM deposition (specific to ECM microcontact printing method)

i. Remove serum-free media from the filter well insert top chamber and replace it with the solution of 10 µg/ml fibronectin in complete growth medium (we have in the past successfully used a solution of 30 µg/ml collagen type I in complete growth medium also for this step).

ii. Incubate the cells in the medium for 30 min at 37 °C.

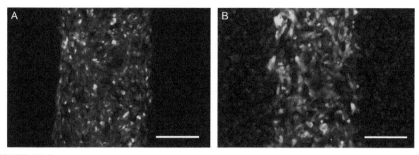

FIGURE 12.2

Example results from each cell patterning stage. (A) Example results of first cell seeding stage. Cells seeded on prepatterned filter inserts should appear confluent once unadhered cells are removed. (B) Example results of second cell seeding stage. Both cells types (first GFP-ARPE-19: green and blue; second wild-type ARPE-19: blue only) should appear confluent once unadhered cells are removed. Scale bar is 100 µm wide. In the example demonstrated here cells were patterned using the parafilm patterning method. (See color plate.)

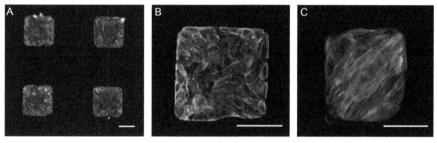

FIGURE 12.3

Cell patterning using agarose microwells method. (A) An array of square shape cell colonies, day 1. (B) Close up view of a square shape cell colony, day 1. (C) Close up view of a square shape cell colony after 14 days in ALI culture. Scale bar is 100 µm wide; ARPE-19 cells were used in this example. (See color plate.)

iii. Remove the solution immediately before seeding the second cell type taking care to prevent drying of the surface since this will be detrimental to the patterned population of the first cell type.

12.2.4.2 Stencil removal (specific to parafilm patterning method)
i. Using forceps separate the corner of the parafilm insert from filter well insert membrane. Gently and steadily peel the parafilm off. The removal works best if parafilm is peeled off as opposed to lifted straight up.
ii. Discard the parafilm insert.

   iii. Inspect under a microscope to ensure that the areas previously under parafilm are free of cells. A clear and confluent pattern of cells should be visible at this stage.

   iv. Next remove the serum-free medium and replace it with the solution of 10 μg/ml fibronectin in complete growth medium (we have in the past successfully used a solution of 30 μg/ml collagen type I in complete growth medium also for this step). Incubate the cells in the medium for 30 min at 37 °C. Remove the solution immediately before seeding the second cell type taking care to prevent drying of the surface.

12.2.5 Seeding of the second cell type (specific to ECM microcontact printing and parafilm patterning coculture methods)

   i. Trypsinize cells to remove them from their culture surface into a solution of medium.

   ii. Centrifuge the cell suspension and resuspend the cells in serum-free medium at a concentration of 2×10^6 cells/ml. Ensure there are no cell aggregates in the resuspended cell solution for better patterning outcomes.

   iii. Apply the cell suspension onto the filter well inserts containing the patterned first cell type. For a six-well insert we use a total of 1 ml per insert, resulting in 1×10^6 cells in a filter insert well. The success of cell patterning is very sensitive to this cell seeding number. If a filter well insert of a different size is used, the cell number should be readjusted based on the surface area of the insert membrane (keeping the cell seeding density at around 2200 cells/mm^2).

   iv. Allow the cells to settle (around 3–5 min) and visually ensure that 100% of the membrane surface is covered by cells. Without disrupting the cells move the filter well insert into an incubator and incubate for 45 min.

   v. Next remove unadhered cells and gently wash the patterns with medium two times.

   vi. Place appropriate volumes of complete growth medium in both chambers of the filter well insert system. If two cell types used in coculture are visually distinguishable from each other (either differentially fluorescently labeled or morphologically different) generated patterns should be clearly visible at this stage. If patterning is successful, you should see confluent cell patterns of cell type 1 surrounded by confluent sheets of cell type 2 (Fig. 12.2B).

12.2.6 Initiation of ALI culture

ALI culture can be initiated for any of the described methods at any time point. It is however important that the agarose microwells, parafilm holes, or the entire surface of filter well insert membrane (in cocultures) are completely covered by cells. Otherwise, the media from the bottom chamber will penetrate into the top chamber disrupting the ALI culture. To set up the ALI culture, simply remove the medium from the top chamber of the filter well insert. Both agarose and parafilm should be left in place for patterned monocultures. Continue the culture for a desired period of time periodically changing the medium in the bottom chamber of the filter well insert and removing any liquid from the top chamber of the insert.

12.2.6.1 Maintaining a prolonged patterned monoculture (specific to agarose microwells method)

We have successfully used this method to maintain patterned cocultures of epithelial cell lines for up to 15 days without any signs of pattern degradation. We foresee that the time at which patterns start degrading will vary based on the cell type used. To ensure maximum patterns stability it is important to prevent agarose from drying. We also suggest taking care not to touch the agarose with pipettes tips or glass pipettes when changing cell culture medium.

12.2.6.2 Maintaining a prolonged patterned monoculture (specific to parafilm patterning method for monocultures)

The stability of parafilm patterning of monocultures in our hands varied greatly from cell to cell type. Certain cell types, such as Madin Darby canine kidney (MDCK) cells, were able to disrupt the imposed pattern overtime by breaking the parafilm–transwell membrane bond and growing under the parafilm insert (Fig. 12.4A). Other cell types, such as ARPE-19, maintained high fidelity to the imposed pattern even 15 days after culture (Fig. 12.4B). We, therefore, suggest that investigators using this method evaluate the pattern stability for their specific cell type. Altering the timing of the annealing step to generate a bond sufficiently strong for the cell type of interest is a possible parameter that could be further optimized to address this issue.

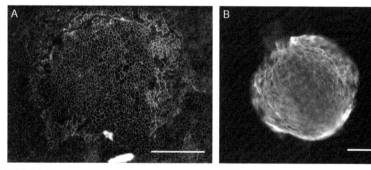

FIGURE 12.4

Prolonged culture of cells patterned using parafilm patterning method. (A) MDCK cells were able to grow under the parafilm insert disrupting the circular colony shape they were originally patterned within. (B) ARPE-19 cells remain in the circular colony shape even after 14 days of ALI culture. Scale bar is 100 μm wide.

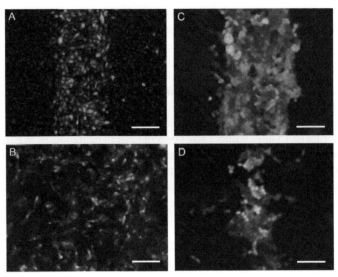

FIGURE 12.5

Pattern stability of cocultures depends on cell type used. (A) Patterned coculture of GFP-ARPE-19 and wild-type ARPE-19 cells on day 1. (B) Patterned coculture of GFP-ARPE-19 and wild-type ARPE-19 cells on day 14. (C) Patterned coculture of GFP-MDCK and wild-type MDCK cells on day 1. (D) Patterned coculture of GFP-MDCK and wild-type MDCK cells on day 14. Scale bar is 100 μm wide; GFP: green, DAPI: blue. (See color plate.)

12.2.6.3 Maintaining a prolonged patterned coculture (specific to ECM microcontact printing and parafilm patterning methods)

In case of cocultures pattern stability greatly depends on the interaction between two cell types used in patterning as well as the migratory behavior of cells. When we pattern MDCK cells, the pattern disruption is minimal (Fig. 12.5A) as these cells are not highly migratory. ARPE-19 cells, on the other hand, are highly migratory and as such quickly deviate from the initially imposed pattern (Fig. 12.5B). The extent of pattern disruption also depends on the pattern feature size (Javaherian, Anesiadis, Mahadevan, & McGuigan, 2013). The acceptable level of pattern disruption is highly dependent on the specific application. We suggest designing the final feature size of patterns such that it accommodates the acceptable level of pattern disruption (Javaherian et al., 2013).

12.3 DISCUSSION

When patterning epithelial cells it is important to accommodate for the ALI culture to induce apical–basal polarization of cells. The ALI culture system requires the use of permeable substrate to allow diffusion of nutrients through the basal side of the cells. The permeable membranes used for ALI culture are usually manufactured from polyester or polycarbonate and contain 0.4–3 μm diameter pores. The three methods presented earlier have been specifically devised to allow the generation of spatially patterned monocultures or cocultures of epithelial cells that can also be exposed to ALI culture. The results presented earlier have been obtained using polyester inserts with 0.4 μm pores; we have also successfully patterned cells using the ECM micro-contact printing method on polycarbonate membranes using the protocol outlined earlier. We, therefore, do not anticipate a need to significantly optimize the presented methods for use with other commercially available filter well inserts.

With all three methods, the success of the patterning is largely dependent on the prepatterning step of the procedure. Accordingly, we will discuss in more detail below the considerations that should be taken to ensure highest quality of prepatterning. Furthermore, each method presented is optimal for specific applications and has its own advantages and limitations, which will also be discussed below.

12.3.1 Agarose microwells method

This method is designed for generating microcolonies of desired shape and size on filter well inserts. Agarose is a cell-repellent hydrogel and is therefore widely used for rendering cell culture substrates nonadhesive. We pattern cells in agarose micro-wells molded using PDMS stamps. Since PDMS stamps used in this method are gen-erated using soft lithography, virtually any shape or size of colony can be generated. Conveniently, once produced, PDMS stamps can be reused multiple times if cleaned and maintained properly. To ensure the best quality of microwells, we ensure full contact between the stamp and the filter insert membrane. This is easily achieved by applying gentle pressure on the stamp after it is placed on the membrane. More-over, we place the filter well insert into an empty well of a multiwell plate when molding agarose. This ensures that the membrane is suspended in the air and agarose cannot perfuse to the bottom side of the membrane. Another important consideration when molding agarose is that it is important to ensure that the agarose is fully gelled before removing the stamp. We suggest placing a droplet of agarose on the corner of the membrane and remove the stamp only after the droplet has completely gelled. Additionally, the removal of the stamp should be carried on by lifting the stamp straight up as opposed to peeling it off. In our hands, cells patterned using agarose microwells on filter well inserts are maintained in patterns for a considerably longer period than on glass or polystyrene substrates. We think that the increased stability of agarose microwells on the filter well inserts results from agarose perfusing the pores of the membrane producing an interlocking effect and hence a stronger bond between agarose and filter well insert than is achieved on flat substrates. Therefore, our method can be used in instances where prolonged culture period is required for none-pithelial cells. In this case, culture medium should be supplied to both chambers of the filter well insert system.

12.3.2 Parafilm patterning method

This method can be used to generate both monocultures and cocultures of epithelial cells. The patterns that can be generated using this method are limited to those that can be carved out of parafilm using blunt needles and scalpel blades. Additionally, the accuracy of generated colony shapes is significantly lower than that of the other two methods as the holes in parafilm are cut out manually. Despite its limitations, however, parafilm method is extremely simple and does not require access to specialized equipment and a clean room. This method, therefore, is ideally suited for use in biology labs that do not possess expertize in microfabrication. Additionally, the cost of cell patterning using parafilm inserts is significantly lower than those of soft lithography-based methods. Furthermore, we have found a new researcher can rapidly optimize this method to achieve reproducible results.

The critical step of prepatterning filter well insert membranes with parafilm stencils is ensuring adequate bonding between parafilm and the transwell membrane. The best result is obtained when the filter well insert is placed on a glass surface and significant pressure is applied from the top to promote bonding between the parafilm and porous membrane. It is critical to apply the pressure vertically without stretching the parafilm. Parafilm can also be easily deformed during the heating step. It is therefore important to ensure that the duration of the heating step is minimal and that the temperature is not high enough to melt the parafilm insert. Conveniently, the quality of the bonding can be evaluated in the degassing step. If the parafilm insert remains bonded to the porous membrane throughout the degassing step, it is highly unlikely that the bond will be compromised at any further step.

Since a significant portion of prepatterning in this method is carried out in nonsterile conditions, it is important to properly sterilize the prepatterned filter well insert before using it. We do so by UV sterilizing for at least 30 min. Ethanol sterilization is not recommended as it weakens the bonding between the parafilm insert and the porous membrane.

12.3.3 ECM microcontact printing patterning method

This method is designed for generation of cocultures. The success of this method is largely dictated by the quality of microcontact printing of the ECM protein onto the porous membrane of filter well insert. To improve the outcome of this step we suggest always preparing a fresh solution of the ECM protein of choice. Next, when removing the ECM solution from the stamps surface, it is important not to touch the surface of the stamp with the KimWipe since this will result in partial removal of ECM from the stamp's surface. If the stamp is not dried properly before placing it on the porous membrane, the remnants of ECM solution will ink the membrane compromising the quality of prepatterning. Finally, it is important to ensure conformal contact between the stamp and the membrane since the protein will not be transferred from any part of the stamp that is not in contact with the membrane. We place 6-g weights on top of the stamp to ensure optimal and even contact; however, it is possible that other filter well insert membranes might require using different weight. It is also critical not to move the stamp once it has come into contact with the surface of membrane. Finally, since the stamps can be reused multiple times for this method, their proper maintenance and cleaning will result in better patterning outcomes over time. Place the stamps in water immediately after using them without allowing them to dry. Clean the stamps by sonication in water supplemented with detergent. Finally, a pressurized stream of ethanol can be used to rinse the stamps after sonication.

12.3.4 The importance of cell seeding parameters

The outcome of patterning is very sensitive to cell seeding parameters. This is especially true for coculture patterning. Two aspects of cell seeding should be optimized to ensure best outcome of patterning: cell seeding density and seeding incubation time. Cell seeding density should be such that when cells sediment, 100% of the surface is covered by cells. For many cells that we have patterned in the past, if the cell density is higher than the optimal density, we have observed cell aggregation, which in turn results in compromised patterning. On the other hand if the cell density is too low, the outcome can also be negatively affected. In case of monocultures, the outcome is not as significantly affected. The resulting cell sheets in agarose microwells or parafilm holes will not be confluent in this case. If this happens, enough time should be allowed for the cell colonies to reach full confluency before ALI culture is imposed. This will introduce variation in the levels of cell confluence in each microcolony however, which may be undesirable for some applications. When patterning cocultures, the low seeding density of first cell type can be detrimental to the outcome. Nonconfluent areas of the first cell type will leave room for the second cell type to attach in areas unintended for them.

The other critical parameter in cell seeding is the incubation time allowed for the first cell type to adhere. Incubation time should be optimized to provide just enough time for adhesion and spreading of cells before unadhered cells are removed. Insufficient time will result in subconfluent cell sheets. Leaving the cells on prepatterned surface for too long may result in cell aggregate formation and/or undesirable adhesion of cells to some of the BSA treated regions of the surface.

12.3.5 Compatibility with ALI culture

The purpose of patterning cells on porous membranes is to be able to impose ALI culture and induce apical–basal polarization. When developing these patterning methods, we considered the possibility that patterning could affect the ability of cells to polarize. To address this concern, we have shown that patterned cells using all three methods polarized to the same extent and with similar timing to nonpatterned cells (Paz et al., 2011, 2012). It is however possible that different cell types may respond differently to patterning and the process of polarization could be affected. Accordingly, we suggest verifying the ability of cells to polarize in patterned cultures, before using these micropatterning methods to carry out experiments using new cell types and combinations.

SUMMARY

Above, we have provided a detailed description of three methods for patterning monocultures and cocultures of epithelial cells. The common feature of all these methods is their compatibility with ALI culture system, which is indispensible for culturing mature polarized epithelium *in vitro*. Agarose microwells patterning methods only allow patterning epithelial monocultures. The main advantages of this method include (i) the ability to generate colonies of virtually any shape and geometry and (ii) pattern stability over prolonged periods of time. The parafilm patterning method can be used to pattern both monocultures and cocultures of epithelial cells. While this method only allows generation of colonies of limited geometries and has a precision lower than that of other two methods, its main attractiveness is its accessibility to virtually any research labs. Finally, the ECM microcontact printing method is designed to pattern cocultures of epithelial cells. Because the method is photolithography based, any colony shape and geometry is possible. It is our hope that by making these micropatterning methods widely accessible they will be used by epithelial biology labs to investigate the effect of colony geometry on variable aspects of epithelium as well as to investigate the importance of cell–cell interactions for epithelium function.

References

Fanto, M., & McNeill, H. (2004). Planar polarity from flies to vertebrates. *Journal of Cell Science*, *117*(Pt 4), 527–533. http://dx.doi.org/10.1242/jcs.00973.

Folch, A., Ayon, A., Hurtado, O., Schmidt, M. A., & Toner, M. (1999). Molding of deep polydimethylsiloxane microstructures for microfluidics and biological applications. *Journal of Biomechanical Engineering*, *121*(1), 28–34.

Frank, S. A. (2007). *Dynamics of cancer: Incidence, inheritance, and evolution* ((2010/09/08 ed.)). Princeton: Princeton University Press.

Gibson, M. C., & Perrimon, N. (2003). Apicobasal polarization: Epithelial form and function. *Current Opinion in Cell Biology*, *15*(6), 747–752.

Gomez, E. W., Chen, Q. K., Gjorevski, N., & Nelson, C. M. (2010). Tissue geometry patterns epithelial-mesenchymal transition via intercellular mechanotransduction. *Journal of Cellular Biochemistry*, *110*(1), 44–51. http://dx.doi.org/10.1002/jcb.22545.

Janvier, R., Sourla, A., Koutsilieris, M., & Doillon, C. J. (1997). Stromal fibroblasts are required for PC-3 human prostate cancer cells to produce capillary-like formation of endothelial cells in a three-dimensional co-culture system. *Anticancer Research*, *17*(3A), 1551–1557.

Javaherian, S., Anesiadis, N., Mahadevan, R., & McGuigan, A. P. (2013). Design principles for generating robust gene expression patterns in dynamic engineered tissues. *Integrative Biology: Quantitative Biosciences from Nano to Macro*, *5*(3), 578–589. http://dx.doi.org/10.1039/c3ib20274g.

Javaherian, S., O'Donnell, K. A., & McGuigan, A. P. (2011). A fast and accessible methodology for micro-patterning cells on standard culture substrates using Parafilm™ inserts. *PLos One*, *6*(6), e20909.

Kaplan, N. A., Liu, X., & Tolwinski, N. S. (2009). Epithelial polarity: Interactions between junctions and apical–basal machinery. *Genetics*, *183*(3), 897–904. http://dx.doi.org/10.1534/genetics.109.108878.

Lang, S. H., Smith, J., Hyde, C., Macintosh, C., Stower, M., & Maitland, N. J. (2006). Differentiation of prostate epithelial cell cultures by matrigel/stromal cell glandular reconstruction. *In vitro Cellular and Developmental Biology. Animal*, *42*(8–9), 273–280. http://dx.doi.org/10.1290/0511080.1.

Nelson, C. M., Liu, W. F., & Chen, C. S. (2007). Manipulation of cell-cell adhesion using bowtie-shaped microwells. *Methods in Molecular Biology*, *370*, 1–10. http://dx.doi.org/10.1007/978-1-59745-353-0_1.

Paz, A. C., Javaherian, S., & McGuigan, A. P. (2011). Tools for micropatterning epithelial cells into microcolonies on transwell filter substrates. *Lab on a Chip*, *11*(20), 3440–3448. http://dx.doi.org/10.1039/c1lc20506d.

Paz, A. C., Javaherian, S., & McGuigan, A. P. (2012). Micropatterning co-cultures of epithelial cells on filter insert substrates. *Journal of Epithelial Biology and Pharmacology*, *5*, 77–85.

Soleas, J. P., Paz, A., Marcus, P., McGuigan, A., & Waddell, T. K. (2012). Engineering airway epithelium. *Journal of Biomedicine and Biotechnology*, *2012*, 982971. http://dx.doi.org/10.1155/2012/982971.

St Johnston, D., & Sanson, B. (2011). Epithelial polarity and morphogenesis. *Current Opinion in Cell Biology*, *23*(5), 540–546. http://dx.doi.org/10.1016/j.ceb.2011.07.005.

Tepass, U. (2012). The apical polarity protein network in Drosophila epithelial cells: Regulation of polarity, junctions, morphogenesis, cell growth, and survival. *Annual Review of Cell and Developmental Biology*, *28*, 655–685. http://dx.doi.org/10.1146/annurev-cellbio-092910-154033.

Tepass, U., & Knust, E. (1993). Crumbs and stardust act in a genetic pathway that controls the organization of epithelia in Drosophila melanogaster. *Developmental Biology*, *159*(1), 311–326. http://dx.doi.org/10.1006/dbio.1993.1243.

Whitesides, G. M., Ostuni, E., Takayama, S., Jiang, X., & Ingber, D. E. (2001). Soft lithography in biology and biochemistry. *Annual Review of Biomedical Engineering*, *3*, 335–373. http://dx.doi.org/10.1146/annurev.bioeng.3.1.335.

Microfabrication of a Platform to Measure and Manipulate the Mechanics of Engineered Microtissues

13

Alexandre Ramade*, Wesley R. Legant†, Catherine Picart*, Christopher S. Chen‡, and Thomas Boudou*

*Laboratory of Materials and Physical Engineering, CNRS UMR5628, and Grenoble Institute of Technology, Grenoble, France
†HHMI Janelia Farm Research Campus, Ashburn, Virginia, USA
‡Department of Bioengineering, University of Pennsylvania, Philadelphia, Pennsylvania, USA

CHAPTER OUTLINE

Methods in Cell Biology, Volume 121
ISSN 0091-679X
http://dx.doi.org/10.1016/B978-0-12-800281-0.00013-0

Abstract

Engineered tissues can be used to understand fundamental features of biology, develop organotypic *in vitro* model systems, and as engineered tissue constructs for replacing damaged tissue *in vivo*. However, a key limitation is an inability to test the wide range of parameters that might impact the engineered tissue in a high-throughput manner and in an environment that mimics the three-dimensional (3D) native architecture. We developed a microfabricated platform to generate arrays of microtissues embedded within 3D micropatterned matrices.

Microcantilevers simultaneously constrain microtissue formation and report forces generated by the microtissues in real time, opening the possibility to use high-throughput, low-volume screening for studies on engineered tissues. Thanks to the micrometer scale of the microtissues, this platform is also suitable for high-throughput monitoring of drug-induced effect on architecture and contractility in engineered tissues. Moreover, independent variations of the mechanical stiffness of the cantilevers and collagen matrix allow the measurement and manipulation of the mechanics of the microtissues. Thus, our approach will likely provide valuable opportunities to elucidate how biomechanical, electrical, biochemical, and genetic/epigenetic cues modulate the formation and maturation of 3D engineered tissues.

In this chapter, we describe the microfabrication, preparation, and experimental use of such microfabricated tissue gauges.

INTRODUCTION

Although two-dimensional (2D) culture models permit some control over substrate stiffness and biochemistry, they nevertheless fail to recapitulate the mechanical and structural environment of native tissue. Thus, potential effects of the extracellular matrix (ECM) protein type, substrate stiffness, and 3D geometry are currently poorly understood. Also, although explanted tissue models retain a more physiological 3D environment, the cell and matrix components cannot be easily controlled and the ability to mimic chronic remodeling events is somewhat restricted.

Over the years, centimeter scale techniques have been developed to encapsulate cells in hydrogels as engineered 3D model systems (Bell, Ivarsson, & Merrill, 1979; Eschenhagen & Zimmermann, 2005; Stopak & Harris, 1982; Yamada & Cukierman, 2007). However, although such systems present a more physiological environment than 2D cultures (Griffith & Swartz, 2006) and may be important for future applications to replace damaged tissues (Prakash & Stenmark, 2012; Zimmermann et al., 2006), they still present important limitations. The scale of these encapsulations (i) necessitates large amounts of cells, a key limitation for delicate cell types (e.g., primary cells, embryonic or induced pluripotent stem cells); (ii) requires histological sectioning to visualize fine scale cellular structures, which is incompatible with live cell microscopy; and (iii) coupled with the diffusion limitations of oxygen and nutrients, as well as drugs and fluorescent dyes, creates unwanted gradients in cell behavior and limits long-term use and drug screening applications. Finally, the cellular forces within these encapsulations are usually inferred only indirectly by measuring the volumetric contraction of free-floating constructs or from constructs pulling against rigid strain gauges (Hansen et al., 2010; Kolodney & Wysolmerski, 1992; Lee, Tsai, Wang, & Ingber, 1998). To obtain quantitative, spatially resolved measurements of cellular forces, investigators have developed 2D soft substrates linked with ECM matrices, such as elastic hydrogels (Butler, Tolić-Nørrelykke, Fabry, & Fredberg, 2002; Dembo & Wang, 1999; Yang, Lin, Chen, & Wang, 2006) and arrays of microfabricated cantilevers (Du Roure et al., 2005; Tan et al., 2003). These force measurement techniques have demonstrated that such forces not only guide mechanical and structural events but also trigger signaling pathways that promote cellular processes such as adhesion, proliferation, or stem-cell differentiation (Engler, Sen, Sweeney, & Discher, 2006; Huang, Chen, & Ingber, 1998; McBeath, Pirone, Nelson, Bhadriraju, & Chen, 2004; Nelson et al., 2005; Parsons, Horwitz, & Schwartz, 2010). However, although these approaches were critical in identifying the important connection between cell mechanics and cell phenotype, the 2D nature of such techniques inherently limits the extent to which 3D morphogenetic phenomena can be described.

We recently developed a microfabricated platform to generate microscale constructs of cells embedded within 3D matrices (Legant et al., 2009). These microfabricated tissue gauges (µTUGs) incorporate microelectromechanical systems (MEMS) cantilevers, which simultaneously constrain and report forces generated by micropatterned 3D constructs in real time (Fig. 13.1). Here we present the simplest protocol which has worked for most cell types we have tested so far, keeping them constrained up to several weeks.

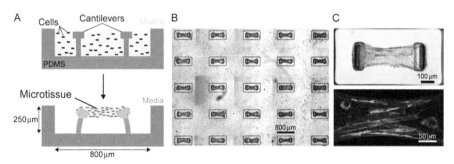

FIGURE 13.1

Generation of microtissues in µTUGs. (A) Schematic illustrating a µTUG and the formation of a microtissue around flexible cantilevers. (B) Large arrays of microtissues simultaneously generated on a substrate. (C) Brightfield image and immunostaining of actin (red), troponin-T (green), and nuclei (blue) in a microtissue composed of C2C12 myoblasts after 10 days of culture. (See color plate.)

This system allows simultaneous monitoring of matrix remodeling events and microtissue force generation, while reporting rapid changes in microtissue force in response to biochemical or physical stimuli. The mechanical stiffness of both the cantilevers and matrix can be easily tuned and electrodes can be added for studying the effect of electrical stimulation on muscle tissues (Boudou et al., 2012).

Thus, this platform represents a unique approach to quantitatively demonstrate the impact of physical parameters on the maturation, structure, and function of 3D microtissues and open the possibility to use high-throughput, low-volume screening for studies on engineered tissues (Boudou et al., 2012; Legant et al., 2009; West et al., 2013; Zhao, Boudou, Wang, Chen, & Reich, 2013).

13.1 PHOTOLITHOGRAPHY REQUIREMENTS

While we present below a brief introduction to photolithography as it pertains to fabricating μTUGs, there are several excellent sources that describe this process in greater details (Campbell, 2001; Jaeger, 2002; Madou, 1997). Standard photolithography involves the exposure of photosensitive material through a photomask.

In our case, the photosensitive material is a negative photoresist (SU-8), meaning that light enables cross-linking of the polymer, making it insoluble to the organic developer. Due to its high viscosity, SU-8 can be spun as thick as 2 mm and can achieve features with height-to-width aspect ratios as great as 25 with standard photoexposure systems.

Photomasks contain microscopic features that are designed like blueprints with a computer-aided design (CAD) tool. A photomask consists of a soda lime glass or quartz plate with a patterned chromium layer, which absorbs visible to deep UV light and prevents exposure to patterned regions of the underlying photoresist.

13.1.1 Materials

Software. A common file format used by manufacturers is GDS II (other formats are CIF and DXF). Any software programs which can produce such a file will work after a short learning phase, either they are free (e.g., KLayout, Glade) or not (e.g., L-Edit, Clewin, AutoCAD).

Photomask. Such masks can be readily obtained from commercial outsourcing services or at local university facilities. Here are three examples of companies producing photomasks: Delta Mask (The Netherlands), Toppan photomasks (present in many countries), and Microtronics Photomasks (USA). Once a design is laid out and fabricated, a single mask can be used repeatedly to generate the same pattern on as many silicon wafers as needed. Another way of designing photomasks consists in sticking printed transparencies on chrome-coated glass blanks, thus decreasing the resolution and the cost.

13.1.2 **Designing geometries**

The basic design of our μTUGs is based on two T-shape cantilevers standing in a rectangular microwell. The presence of a wide cap at the tip of the cantilevers is critical to constrain the microtissue formation. To achieve this complex geometry, μTUGs are fabricated by using successive spin coat, alignment, exposure, and baking steps, in order to create multilayer templates (Fig. 13.2). The first layer, or base coating, ensures a good bonding of the cantilevers to the silicon wafer. The second one, or bottom layer, allows the creation of the posts. A third thin interstitial layer, which strongly absorbed UV light, serves as a lithographic-stop layer and prevents unwanted cross-linking of the underlying SU-8 layers. The fourth layer, or top layer, leads to the creation of the top wide cap at the tip of the post.

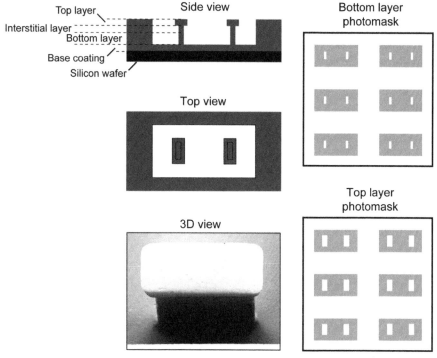

FIGURE 13.2

Schematic of the μTUGs and the photomasks. Left column: side view, top view and representative SEM image illustrating the multilayered structure of the μTUGs. The sample is tilted 30° to simultaneously visualize the wide cap at the tip of the post and the base. Right column: schematic of the bottom and the top layer photomasks necessary to fabricate μTUGs. (For color version of this figure, the reader is referred to the online version of this chapter.)

Thus, two photomask designs are necessary to fabricate µTUGs. The first one, the bottom layer photomask, serves to create the bottom part of the cantilevers whereas the second one, the top layer photomask, serves to shape the wide cap at the top of the cantilevers (Fig. 13.2).

13.1.3 Characterization of cantilever spring constant *k*

The µTUGs are force sensors, composed of a transparent silicone rubber called poly-dimethylsiloxane (PDMS). Each cantilever deflection is proportional to its spring constant *k* and reports the traction force generated by the microtissue. The spring constant can be estimated through classical relationships describing beam bending. A cantilever can be regarded as a beam, fixed at one end and loaded with forces at the other end, undergoing pure bending and negligible shearing. From the theory of slender beam bending, the force *F* can thus be calculated from the geometry of the cantilever (the width *w*, the thickness *t*, the length *L*, and the length from bottom to centroid of the cap *a*), its deflection δ and the Young's modulus of the PDMS *E* (Fig. 13.3). The Young's modulus of PDMS (0.5–5 MPa, depending on the ratio PDMS/curing agent) can be quantified via uniaxial tensile testing on PDMS strips. The thickness, width, and length of the cantilevers are measured optically. Calculation of this relationship provides the conversion from tip displacement to applied force.

The spring constant of the posts can also be calibrated experimentally, either by using calibrated glass micropipettes (Legant et al., 2009) or capacitive MEMS force sensor (Boudou et al., 2012). We generally find close agreement between these direct measurements of *k* and calculated *k* as described earlier based on properties of PDMS and dimension of the cantilever.

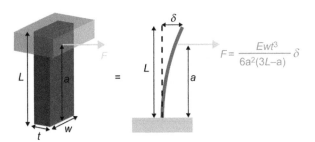

$$F = \frac{Ewt^3}{6a^2(3L-a)}\delta$$

FIGURE 13.3

Characterization of cantilever mechanics: cantilever deflection is modeled as 1D beam bending due to a load applied at the centroid of the top cap. The force is then calculated from the thickness *t*, the width *w*, the total length *L*, the length from bottom to centroid of the cap *a*, the deflection δ, and the Young's modulus of the PDMS *E*. (For color version of this figure, the reader is referred to the online version of this chapter.)

13.2 **MICROFABRICATION OF THE PLATFORM**

Although we describe the fabrication steps in detail, results will vary depending on the equipment used. We strongly suggest that these process parameters be optimized according to specific facilities.

13.2.1 **Microfabrication of the master mold (Fig. 13.4)**

13.2.1.1 *Materials*

- Silicon wafer, 100 mm (e.g., MEMC Electronic Materials, Silicon Quest International)
- Negative photoresists SU8 2002, 2010, and 2100 (Microchem)
- Positive photoresist S1813 (Microchem)
- Propylene glycol monomethyl ether acetate (PGMEA, Sigma)
- Isopropyl alcohol (IPA, Sigma)

Caution: Photoresists and PGMEA need to be handled with care. Wear chemical goggles, chemical gloves, and suitable protective clothing. Do not get into eyes, or onto skin or clothing. Use with adequate ventilation to avoid breathing vapors or mist.

13.2.1.2 *Equipments*

- Spin coater (e.g., model WS-400-6NPP-LITE, Laurell Technologies Corporation)
- UV-ozone cleaner (e.g., Model 342, Jelight)
- Mask aligner (e.g., model MJB3, Suss Microtec) using a U-360 band pass filter (Hoya Optics)
- Oven and/or hot plate
- Luxmeter

13.2.1.3 *Method*

1. Preparing the wafer:
   a. Dehydrate wafers at 250 °C for 30 min.
   b. UV-ozone for 15 min to prime the wafer surface for optimal attachment of SU-8.
2. Base coating and exposure:
   a. Spincoat SU8 2002 on the silicon wafer for 10 s at 500 rpm (acceleration 84 rpm/s), then 60 s at 2000 rpm (acc. 252 rpm/s).
   b. Soft bake 2 min at 95 °C on hot plate to evaporate out solvent from the polymer film.
   c. Expose to UV at 100 mJ of total light energy (=power × time).
   d. Postexposure bake 2 min at 95 °C to thermally drive the cross-linking reaction.
   e. Take wafer off hot plate and slowly cool to room temperature to reduce cracking in the SU-8 film.
3. Bottom and interstitial layers coating:

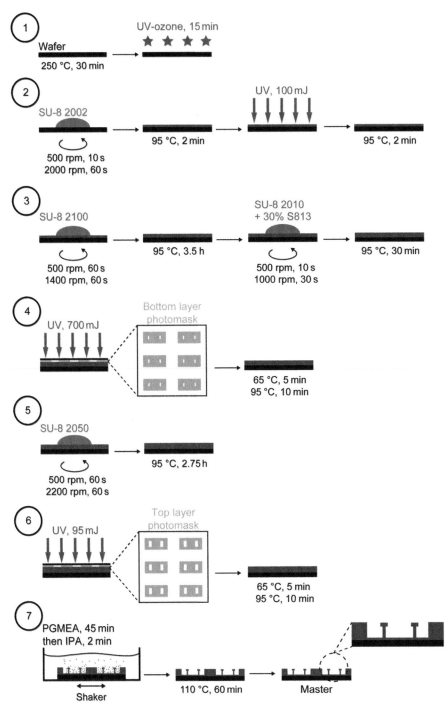

FIGURE 13.4

Microfabrication of the master wafer. After dehydration and cleaning of the wafer (1), four SU-8 layers are successively spin coated, exposed, and baked (2–6) before development in PGMEA and final bake (7). The obtained master wafer contains microwells containing T-shape cantilevers. (For color version of this figure, the reader is referred to the online version of this chapter.)

   a. Spincoat SU8 2100 on the wafer for 60 s at 500 rpm (acc. 84 rpm/s), then 60 s at 1400 rpm (acc. 252 rpm/s).
   b. Soft bake immediately 3.5 h at 95 °C.
   c. Allow hot plate to cool below 50 °C.
   d. Spincoat a premixed solution of doped SU8 2010+30% S1813 for 10 s at 500 rpm (acc. 84 rpm/s), then 30 s at 1000 rpm (acc. 252 rpm/s).
   e. Soft bake immediately 30 min at 95 °C.
4. Bottom and interstitial layers exposure:
   a. Place the bottom layer chrome mask on the wafer and expose to UV (700 mJ).
   b. Postexposure bake 5 min at 65 °C, then 10 min at 95 °C.
5. Top layer coating:
   a. Spincoat SU8 2050 on the wafer for 60 s at 500 rpm (acc. 84 rpm/s), then 60 s at 2200 rpm (acc. 252 rpm/s).
   b. Soft bake immediately 2.75 h at 95 °C.
6. Top layer exposure:
   a. Align the top layer mask so that the top part of the cantilevers (wider) frames the bottom part of the cantilevers (thinner). Expose to UV (95 mJ).
   b. Postexposure bake 5 min at 65 °C, then 10 min at 95 °C.
7. Developing the master:
   a. Immerse the coated master wafer in PGMEA on shaker for 45 min to wash away the soluble regions, leaving behind the insoluble patterns of cross-linked photoresist.
   b. Immerse the master in IPA on shaker for 2 min to rinse off the PGMEA.
   c. Remove the master wafer from the IPA and hard bake for 60 min at 110 °C.

13.2.2 Replication of the master mold (Figs. 13.5 and 13.6)
13.2.2.1 Materials
- Trichloro(1H,1H,2H,2H-perfluorooctyl)silane (Sigma)
- PDMS (Sylgard 184, Dow Corning)
- Razor blades
- IPA (Sigma)
- 35-mm petri dishes (BD Biosciences)
- Optional: Fluorescent microbeads (3 μm diameter, Polysciences #17147)

Caution: Trichloro(1H,1H,2H,2H-perfluorooctyl)silane needs to be handled with care. Wear chemical goggles, chemical gloves, and suitable protective clothing. Do not get into eyes, or onto skin or clothing. Use with adequate ventilation to avoid breathing vapors or mist.

13.2.2.2 Equipments
- Plasma cleaner (e.g., SPI Supplies)
- Desiccator
- Oven
- Shaker

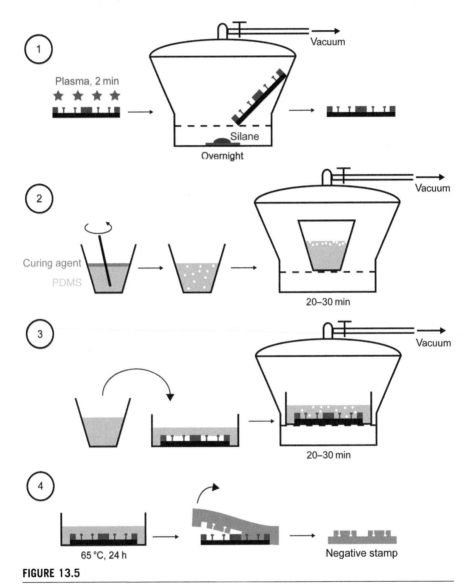

FIGURE 13.5

Replication of the master mold. After silanization (1) the degased mixture of PDMS and curing agent (2) is poured on the master mold before further degasing (3) and de-molding (4). (For color version of this figure, the reader is referred to the online version of this chapter.)

13.2.2.3 First replication (Fig. 13.5)

1. Silanization (to prevent PDMS from permanently bonding to the master mold):
   a. Plasma clean for 2 min.
   b. Pour ~5 drops of trichloro(1H,1H,2H,2H-perfluorooctyl)silane on a glass slide laying on the bottom of a desiccator.

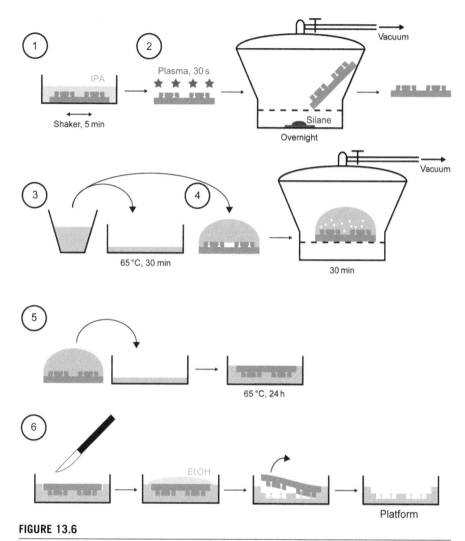

FIGURE 13.6

Replication of the negative PDMS stamp. After cleaning and silanization (1), PDMS is poured in the petri dish (3) and on the negative stamp (4). The negative stamp is then inverted into the petri dish and the PDMS is cured (5) before de-molding. (For color version of this figure, the reader is referred to the online version of this chapter.)

    c. Place the master tilted at 45° in the desiccator, connect the desiccator to vacuum and leave overnight.
2. Preparation of the PDMS (~30 g/master):
    a. In a beaker, pour PDMS and curing agent in a 10:1 ratio.
    b. Mix well with a glass or plastic stick (poor mixing can lead to heterogeneous PDMS).

    c. Place the beaker in the desiccator and apply vacuum for 20–30 min (depending on the amount of PDMS) until no air bubbles are visible and the PDMS is transparent.

3. Molding the PDMS:

    a. Pour a small amount of PDMS on the features of the master.

    b. Introduce the master into the desiccator and apply vacuum for 20–30 min, until no air bubbles are visible.

    c. Pour the remaining amount of PDMS on the master.

    d. Cure at 65 °C for 24 h.

4. De-molding the negative stamp:

    a. Carefully cut through the PDMS around the master wafer with a razor blade.

    b. Very carefully peel the PDMS layer from the silicon wafer. At this step, the obtained PDMS stamp is a negative stamp of the master, which needs to be replicated a second time to obtain the final platform.

13.2.2.4 Second replication (Fig. 13.6)

1. Cleaning the stamp:

    a. Place the stamp with the pattern face up into a small glass dish.

    b. Pour enough IPA to fully submerge the stamp.

    c. Agitate gently for 5 min on a shaker.

    d. Blow dry with N_2 gas.

2. Silanization (to prevent PDMS from permanently bonding to the stamp):

    a. Place the stamp on a glass slide (pattern face up).

    b. Plasma treatment for 45 s to activate the PDMS surface groups.

    c. Pour ~5 drops of trichloro(1H,1H,2H,2H-perfluorooctyl)silane on a glass slide laying on the bottom of a desiccator.

    d. Place the glass slides with the stamps tilted at 45° in the desiccator under vacuum and leave overnight.

3. Molding the PDMS (the next day):

    a. Optional step: Inserting fluorescent microbeads

    This step is not necessary but in order to track precisely the deflection of the cantilevers and facilitate image analysis, the insertion of fluorescent microbeads at the top of the cantilevers is required.

    Put the stamps in a six-well plate, dilute the fluorescent beads at 0.5×10^6 beads/ml in ethanol and cover each stamp with 10 ml of it. Centrifuge at 1000 rpm for 1 min. Let the ethanol evaporate.

    b. Prepare the PDMS (~7–10 g/stamp) as described in Section 13.2.2.3 (step 2) The ratio PDMS/curing agent can be varied to modulate the spring constant of the cantilevers (see Section 13.1.3).

    c. Pour ~1.5 g of PDMS on the bottom of the 35-mm petri dish (enough to cover the entire bottom with approximately 1 mm of PDMS).

    d. Place on a hot plate at 65 °C for 20–30 min (the PDMS becomes slightly stiffer). This step is important to prevent the stamp from sinking all the way to the bottom of the dish in step 5.

4. Preparing the stamp:
  a. While petri dish is on hot plate, remove stamp from desiccator and pour a small amount of PDMS onto the stamp (enough to cover entire pattern).
  b. Degas in the desiccator under high vacuum until no bubbles are visible.
5. Inverting and curing the stamp onto petri dish:
  a. Carefully peel stamp from glass with a fine pair of tweezers.
  b. Invert the stamp (pattern should be going into the petri dish) and lower into petri dish at an angle allowing one side to come in contact first.
  c. Slowly lower the stamp to the completely horizontal position. Watch the interface between the PDMS on the stamp and PDMS on the dish. Lower the stamp slowly to avoid trapping bubbles between the dish and stamp.
  d. Center stamp in the petri dish and add PDMS into the void between the stamp and the side of the dish until PDMS is level with the top of the stamp.
  e. Place substrates into a 65 °C oven for 24 h.
6. De-molding the stamp:
  a. Carefully cut the PDMS at the interface of the stamp and substrate with a razor blade (substrate refers to the petri dish/PDMS complex henceforth). The interface should be visible from a top view.
      i. If little or no PDMS is on the top of the stamp it should be possible to cut very lightly and apply torque to the razor blade so that the PDMS interface comes apart, provided the silanization went well.
      ii. Repeat for each side of the stamp, possibly going around twice making sure the sides of the stamp are detached.
      iii. Take care not to perturb the bottom interface between the pattern on the stamp and the PDMS on the petri dish yet.
  b. Pipette ~200 μl ethanol around the sides of the stamp into the crevice. Make sure that the sides of the stamp are exposed to ethanol. This will make the PDMS less adhesive, facilitating easy peeling.
  c. Peel the stamp away from the substrate. This step is absolutely critical for success and requires practice to achieve satisfactory results. Applying too much force can lead to tearing away the cantilevers, clogging the stamp, and thereby decreasing its quality.
      i. Orient the stamp so that the horizontal portion of the wells is facing away from you. (This will ensure that the cantilevers are oriented vertically which will allow them to resist more stress during the peeling process.)
      ii. Insert a pair of tweezers all the way down the crevice between the stamp and substrate PDMS along the side farthest from you.
      iii. Apply slight pressure on the tweezers away from yourself. The receding interface between the pattern on the stamp and the substrate should be visible. Do not pry the stamp away at this point. Slowly apply increasing pressure to propagate the detachment.
      iv. Begin gently prying the stamp away. Take extreme care to go as slowly as possible during this step to avoid breaking the cantilevers.
      v. Once the stamp is free, peel any excess PDMS that has adhered to the stamp off.

    vi. Blow dry both using N_2.

    vii. Rinse both the stamp and substrate in IPA again and blow dry. At this step, the obtained PDMS platform should be an exact replicate of the master. Check under the microscope that the wells and cantilevers are well shaped before seeding cells in it.

13.3 CELL SEEDING (FIG. 13.7)

The described μTUGs have been used successfully with the following cells: NIH3T3 fibroblasts, neonatal rat cardiomyocytes (NRCM), cardiomyocytes derived from human-induced pluripotent stem cells, immortalized and primary human airway smooth muscle (ASM) cells, and C2C12 skeletal muscle cells. We are confident that μTUGs can be used with other cell types after optimizing the process parameters, especially the duration of the pluronic treatment (step 2) and the composition of the matrix (step 3).

13.3.1 Materials

- Phosphate buffered saline (PBS) (Invitrogen/Gibco)
- Trypsin (0.25 g/l)–EDTA(0.2 g/l) (Invitrogen/Gibco)
- Culture medium (adapted to the cell type)
- Pluronic F127 (Sigma P2443)
- Liquid neutralized collagen I from rat tail (BD Biosciences)
- M1999 $10\times$ (Sigma)
- NaOH
- $NaHCO_3$
- HEPES (4-(2-hydroxyethyl)-1-piperazineethanesulfonic acid)

13.3.2 Specific equipment

- Centrifugator (e.g., Eppendorf 5702, Dutscher)
- Cooler pack
- Desiccator

13.3.3 Method

Turn on the centrifugator and cool it to a temperature lower than 0 °C.

1. Sterilizing the platforms:
   a. Sterilize under UV for 15 min.
   b. Fill the platform to the top with 70% ethanol/30% water, and pipette up and down with 1 ml pipette to dislodge any trapped air pockets in the molds.
   c. Aspirate off ethanol and blow dry under N_2.

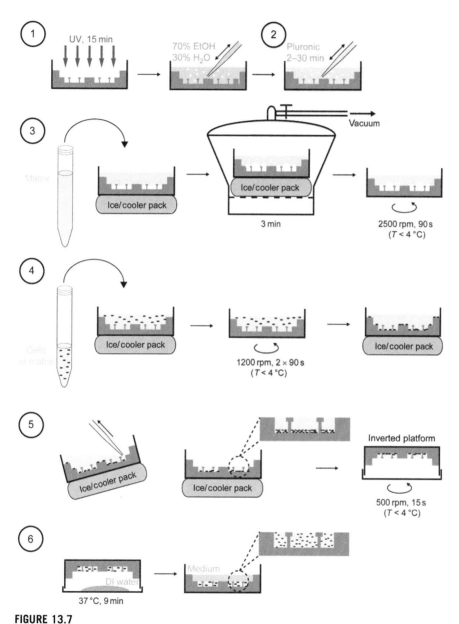

FIGURE 13.7

Cell seeding in μTUGs. After sterilization (1), the platform is treated with pluronic to reduce cell adhesion to the PDMS (2) and a solution of collagen is introduced within the microwells (3). Cell are then resuspended in the collagen solution and centrifuged into the wells (4). Excess collagen is then removed (5) and the remaining collagen polymerized (6). (See color plate.)

2. Pluronic treatment of the molds:
   a. Fill the platform to the top with 0.2% solution of F127 in PBS. Pipette up and down with 1 ml pipette to dislodge any trapped air pockets. The pluronic will reduce the cell adhesion to the PDMS, allowing the forming tissue to detach from the sides of the microwell. Consequently, the duration of the pluronic treatment strongly depends on the cell type and their density. The more the cells are contractile (or the higher their density is), the faster they will compact the matrix and detach from the wall and the shorter the pluronic treatment should be. This duration needs to be optimized for each cell type and density (see Table 13.1 for examples).
   b. Aspirate off pluronic solution and rinse once with PBS. Blow dry under N_2.
3. Preparing collagen solution (always keep the collagen solution and the platforms on ice or on the cooler pack to avoid premature polymerization):
   a. Scale the volumes presented in Table 13.2 to the desired final volume of solution (2 ml/platform with platforms molded in 35-mm petri dishes) and add the ingredients to a tube kept on ice. Collagen gels with a concentration lower than 0.5 mg/ml cannot sustain the compaction exerted by the cells whereas a concentration higher than 2.5 mg/ml can lead to difficulties during the dewetting step (step 5).
   The matrix can be completed with other components such as fibrinogen, matrigel, and hyaluronic acid. Adapt the water volume and pH if necessary.

Table 13.1 Examples of the duration of the pluronic treatment in function of the cell type and density

Cell type	Cell density (cells/well)	Pluronic treatment duration (min)
NIH3T3 fibroblasts	100	5
NIH3T3 fibroblasts	300	2
C2C12 myoblasts	300	2
NRCM	1000	30

Table 13.2 Collagen formulation

	Collagen 1 mg/ml (µl)	Collagen 2 mg/ml (µl)
H_2O	1117	531
M199 10×	200	200
HEPES (250 mM)	80	80
NaOH (1 M)	24	45
$NaHCO_3$ (5% w/v)	14	14
Collagen (3.54 mg/ml)	565	1130

   If the addition of other components changes the polymerization kinetics of the hydrogel, be sure to adapt the polymerization time at step 6b.

   b. Mix everything well. Final solution should be "salmon" colored, or roughly the color of CO_2 equilibrated culture medium. If not, equilibrate the pH to 7.4 with NaOH (1 M) or HCl (1 M) by using a pH meter.

   c. Add 1 ml of the collagen solution to each platform. It should be enough to fully cover the bottom surface of the mold with some excess on top.

   d. Degas the platforms to pull the collagen solution into each well (3 min).

   e. Centrifuge at ~2500 rpm for 1.5 min in a temperature-controlled centrifuge set to $T < 0\,°C$. Check if any bubbles subsist and redo the centrifugation one time if some are still present.

4. Seeding cells:

   a. Spin down appropriate number.
   The final cell density has to be optimized for each cell type. Typical densities vary from 100 cells/well (e.g., fibroblasts, highly proliferative and contractile) to 1000 cells/well (e.g., cardiomyocytes, nonproliferative and weakly contractile).

   b. Aspirate off media from cells, add 0.5 ml of the collagen solution to resuspend the cells.

   c. Add back 0.5 ml cell laden collagen solution to each platform and mix well with the collagen already in the platform.

   d. Spin down the cells into the molds two times at 1200 rpm for 1.5 min in a temperature-controlled centrifuge set to $T < 0\,°C$ (rotating the platforms 180° between the two centrifugations helps obtaining a homogeneous seeding). At this step, all the cells should lay on the bottom of the platform, inside and outside the microwells.

5. Removing excess collagen:

   a. Slightly tilt the platforms on the ice. Starting at a corner of the platform, slowly aspirate off the collagen solution. The collagen should "dewet" in a sheet from the PDMS surface, leaving behind only pockets of cells/collagen in each template. This step is absolutely critical for success and requires practice to achieve satisfactory results. At this step, all the cells should lay on the bottom of the platform, only inside the microwells.

   b. Place the molds inverted in the centrifuge and spin at 500 rpm for 15 s. At this step, the cells should be homogeneously located within the matrix pockets in the microwells.

6. Polymerizing the collagen:

   a. Add 1 ml of sterile diH_2O (prewarmed to 37 °C) to the lid of each mold and invert the mold onto the lid (this keeps the collagen from drying out during the next step).

   b. Place the samples (inverted) into an incubator for 9 min. If some collagen solution remains in the tube, place it into the incubator too, to verify that the collagen solution polymerizes.

   c. Remove the molds from the lids. Starting at a corner of the mold, very slowly add growth media (1–2 ml total) to the platforms.

   d. Replace the lids onto the molds and place in incubator.

13.4 MEASUREMENTS

13.4.1 Image analysis

To calculate the magnitude of the tip deflections, we measure the displacement of the top of the cantilevers using an A-Plan $10 \times$ objective on an inverted microscope. At least two images are necessary to determine the cantilever deflection: the first one is the zero-force reference image (usually taken before cell seeding) and the second one is taken when the microtissue is formed. To analyze dynamic processes, such as twitch contractions of cardiac microtissues, streams of images are necessary. Phase contrast images can be used to measure cantilever displacements using Image J (National Institutes of Health) but images of the fluorescent microbeads will allow better precision and faster analysis. The displacement of the fluorescent microbeads at the top of the cantilevers is tracked by using the SpotTracker plug-in in Image J. The displacement of the top of the cantilevers (in pixels), compared with the reference image, is then converted to traction forces by converting pixel to micrometer and multiplying the displacement by the spring constant of the cantilevers (see Section 13.1.3).

13.4.2 Staining of μTUGs arrays

The microtissues can be fixed and stained to identify specific structures, organelles, or proteins. We use 4% paraformaldehyde in PBS for 30 min as our fixing solution to cross-link proteins and 0.2–0.5% Triton X-100 in PBS for 1 h to permeabilize the membrane. To visualize F-actin, microtissues can be directly stained with fluorophore-conjugated phalloidin (1:200, Invitrogen) and Hoechst nuclear dye (1:1000, Invitrogen). To visualize other specific structures, microtissues need to be blocked after permeabilization, in 5–20% of the secondary antibody host serum in 5% bovine serum albumin (Sigma) in PBS. Microtissues are then incubated with primary antibodies and detected with fluorophore-conjugated, isotype-specific, anti-IgG antibodies. As microtissues are thick samples, we recommend letting primary antibodies overnight at 4 °C to allow homogeneous diffusion.

   The samples can then be mounted onto a glass slide with a microscope cover glass overlaid on top, using antifade reagent (e.g., ProLong Gold, Molecular Probes-Invitrogen), for fluorescent and phase-contrast microscopy at high magnification.

13.5 EXPERIMENTAL APPLICATIONS OF μTUGs AND DISCUSSION

The protocol presented in this chapter is a simple and yet robust protocol to generate microtissues suspended between two cantilevers that allow the quantification of the microtissue contractility. We observed traction forces ranging from 0.1 to 30 μN, depending on the cell type, the cell density, the matrix stiffness, the spring constant of the cantilevers, and the maturation of the microtissue. The lower limit on force resolution depends on the magnification of the microscope and the method of analysis. By introducing fluorescent microbeads at the top of the cantilevers, this limit is ~0.1 μN with a $10\times$ objective and a subpixel analysis of the beads displacement with Image J.

Several simple modifications can be made to the μTUGs in order to study specific processes. First the spring constant of the cantilevers can easily be adjusted by changing the ratio PDMS/curing agent or the cantilever geometry. Through screening a range of PDMS stiffness, we observed that such variations in the spring constant of the cantilevers influence the microtissue contractility (Boudou et al., 2012; Legant et al., 2009). Similarly, the stiffness and biochemistry of the ECM can be modified by changing the ECM protein type and its density, which further leads to changes in the microtissue contractility. We have looked at traction forces with respect to cell types and have found that there is also phenotypic specificity in force generation for NIH3T3 fibroblasts, NRCM, immortalized and primary human ASM cells, and C2C12 skeletal muscle cells (Boudou et al., 2012; Legant et al., 2009; Sakar et al., 2012; West et al., 2013).

Second, the geometry of the microtissue can be changed by modifying the shape and alignment of the cantilevers. This modification allows the study of the influence of the spatial organization of the environment (e.g., two cantilevers leading to uniaxial loading of the microtissue versus four cantilevers leading to multiaxial loading) on the architecture of the microtissue (Legant et al., 2009).

Third, two parallel carbon electrodes can be inserted on both sides of the arrays and thus allow electrical stimulation of arrays of microtissues. By continuously pacing cardiac microtissues, we confirm previous results suggesting that electrical stimulation provides cardiomyocytes with maturation signals (Boudou et al., 2012).

Finally, we recently demonstrated the possibility to couple the μTUGs technology with magnetic actuation (Zhao et al., 2013) and optogenetic control of the microtissues (Sakar et al., 2012), opening exciting avenues for studying fundamental processes or for developing applications such as soft robots.

In conclusion, the overall similarity of structural and functional characteristics between microtissues and *in vivo* tissues is promising and the method described here opens the potential to high-throughput, low-volume screening applications. Moreover, this system permits quantitative assessment of physical parameters on the maturation, structure, and function of physiological and pathological models. Most importantly, this system provides reproducible contractile phenotyping that is virtually absent in 2D culture models. These same attributes will likely provide valuable opportunities to elucidate how biomechanical, electrical, biochemical, and genetic/epigenetic cues modulate the differentiation and maturation of stem cells. Thus, combining stem cell differentiation and microtissue engineering could pave the way to the production of high-quality functional microtissues from stem cells, opening an exciting avenue for the treatment of various damaged tissues.

References

Bell, E., Ivarsson, B., & Merrill, C. (1979). Production of a tissue-like structure by contraction of collagen lattices by human fibroblasts of different proliferative potential in vitro. *Proceedings of the National Academy of Sciences of the United States of America*, *76*(3), 1274–1278.

Boudou, T., Legant, W. R., Mu, A., Borochin, M. A., Thavandiran, N., Radisic, M., et al. (2012). A microfabricated platform to measure and manipulate the mechanics of engineered cardiac microtissues. *Tissue Engineering. Part A*, *18*(9–10), 910–919.

Butler, J. P., Tolić-Nørrelykke, I. M., Fabry, B., & Fredberg, J. J. (2002). Traction fields, moments, and strain energy that cells exert on their surroundings. *American Journal of Physiology. Cell Physiology*, *282*(3), C595–C605.

Campbell, S. (2001). The science and engineering of microelectronic fabrication (The Oxford series in electrical and computer engineering, Ed.) (2nd ed.). (p. 603). New York: Oxford University Press.

Dembo, M., & Wang, Y. L. (1999). Stresses at the cell-to-substrate interface during locomotion of fibroblasts. *Biophysical Journal*, *76*(4), 2307–2316.

Du Roure, O., Saez, A., Buguin, A., Austin, R. H., Chavrier, P., Silberzan, P., et al. (2005). Force mapping in epithelial cell migration. *Proceedings of the National Academy of Sciences of the United States of America*, *102*(7), 2390–2395.

Engler, A. J., Sen, S., Sweeney, H. L., & Discher, D. E. (2006). Matrix elasticity directs stem cell lineage specification. *Cell*, *126*(4), 677–689.

Eschenhagen, T., & Zimmermann, W. H. (2005). Engineering myocardial tissue. *Circulation Research*, *97*(12), 1220–1231.

Griffith, L. G., & Swartz, M. A. (2006). Capturing complex 3D tissue physiology in vitro. *Nature Reviews. Molecular Cell Biology*, *7*(3), 211–224.

Hansen, A., Eder, A., Bönstrup, M., Flato, M., Mewe, M., Schaaf, S., et al. (2010). Development of a drug screening platform based on engineered heart tissue. *Circulation Research*, *107*(1), 35–44.

Huang, S., Chen, C. S., & Ingber, D. E. (1998). Control of cyclin D1, p27(Kip1), and cell cycle progression in human capillary endothelial cells by cell shape and cytoskeletal tension. *Molecular Biology of the Cell*, *9*(11), 3179–3193.

Jaeger, R. (2002). *Introduction to microelectronic fabrication* (2nd ed.). (p. 316). Upper Saddle River, NJ: Prentice Hall.

Kolodney, M. S., & Wysolmerski, R. B. (1992). Isometric contraction by fibroblasts and endothelial cells in tissue culture: A quantitative study. *The Journal of Cell Biology*, *117*(1), 73–82.

Lee, K. M., Tsai, K. Y., Wang, N., & Ingber, D. E. (1998). Extracellular matrix and pulmonary hypertension: Control of vascular smooth muscle cell contractility. *The American Journal of Physiology*, *274*(1 Pt 2), H76–H82.

Legant, W. R., Pathak, A., Yang, M. T., Deshpande, V. S., McMeeking, R. M., & Chen, C. S. (2009). Microfabricated tissue gauges to measure and manipulate forces from 3D micro-tissues. *Proceedings of the National Academy of Sciences of the United States of America, 106*(25), 10097–10102.

Madou, M. (1997). *Fundamentals of microfabrication.* Boca Raton, FL: CRC Press589.

McBeath, R., Pirone, D. M., Nelson, C. M., Bhadriraju, K., & Chen, C. S. (2004). Cell shape, cytoskeletal tension, and RhoA regulate stem cell lineage commitment. *Developmental Cell, 6*(4), 483–495.

Nelson, C. M., Jean, R. P., Tan, J. L., Liu, W. F., Sniadecki, N. J., Spector, A. A., et al. (2005). Emergent patterns of growth controlled by multicellular form and mechanics. *Proceedings of the National Academy of Sciences of the United States of America, 102*(33), 11594–11599.

Parsons, J. T., Horwitz, A. R., & Schwartz, M. A. (2010). Cell adhesion: Integrating cytoskeletal dynamics and cellular tension. *Nature Reviews. Molecular Cell Biology, 11*(9), 633–643.

Prakash, Y. S., & Stenmark, K. R. (2012). Bioengineering the lung: Molecules, materials, matrix, morphology, and mechanics. *American Journal of Physiology. Lung Cellular and Molecular Physiology, 302*(4), L361–L362.

Sakar, M. S., Neal, D., Boudou, T., Borochin, M. A., Li, Y., Weiss, R., et al. (2012). Formation and optogenetic control of engineered 3D skeletal muscle bioactuators. *Lab on a Chip, 12*(23), 4976–4985.

Stopak, D., & Harris, A. K. (1982). Connective tissue morphogenesis by fibroblast traction. *Developmental Biology, 90*(2), 383–398.

Tan, J. L., Tien, J., Pirone, D. M., Gray, D. S., Bhadriraju, K., & Chen, C. S. (2003). Cells lying on a bed of microneedles: An approach to isolate mechanical force. *Proceedings of the National Academy of Sciences of the United States of America, 100*(4), 1484–1489.

West, A. R., Zaman, N., Cole, D. J., Walker, M. J., Legant, W. R., Boudou, T., et al. (2013). Development and characterization of a 3D multicell microtissue culture model of airway smooth muscle. *American Journal of Physiology. Lung Cellular and Molecular Physiology, 304*(1), L4–L16.

Yamada, K. M., & Cukierman, E. (2007). Modeling tissue morphogenesis and cancer in 3D. *Cell, 130*(4), 601–610.

Yang, Z., Lin, J.-S., Chen, J., & Wang, J. H.-C. (2006). Determining substrate displacement and cell traction fields—A new approach. *Journal of Theoretical Biology, 242*(3), 607–616.

Zhao, R., Boudou, T., Wang, W.-G., Chen, C. S., & Reich, D. H. (2013). Decoupling cell and matrix mechanics in engineered microtissues using magnetically actuated microcantilevers. *Advanced Materials, 25*(12), 1699–1705.

Zimmermann, W.-H., Melnychenko, I., Wasmeier, G., Didié, M., Naito, H., Nixdorff, U., et al. (2006). Engineered heart tissue grafts improve systolic and diastolic function in infarcted rat hearts. *Nature Medicine, 12*(4), 452–458.

Methods for Two-Dimensional Cell Confinement

14

Maël Le Berre, Ewa Zlotek-Zlotkiewicz, Daria Bonazzi, Franziska Lautenschlaeger, and Matthieu Piel

Systems Cell Biology of Cell Division and Cell Polarity, Institut Curie, CNRS, Paris, France

CHAPTER OUTLINE

Abstract

Protocols described in this chapter relate to a method to dynamically confine cells in two dimensions with various microenvironments. It can be used to impose on cells a

http://dx.doi.org/10.1016/B978-0-12-800281-0.00014-2

given height, with an accuracy of less than 100 nm on large surfaces (cm^2). The method is based on the gentle application of a modified glass coverslip onto a standard cell culture. Depending on the preparation, this confinement slide can impose on the cells a given geometry but also an environment of controlled stiffness, controlled adhesion, or a more complex environment. An advantage is that the method is compatible with most optical microscopy technologies and molecular biology protocols allowing advanced analysis of confined cells. In this chapter, we first explain the principle and issues of using these slides to confine cells in a controlled geometry and describe their fabrication. Finally, we discuss how the nature of the confinement slide can vary and provide an alternative method to confine cells with gels of controlled rigidity.

INTRODUCTION

Numerous methods have been used to study cells in a confined environment. The most common methods have certainly been three-dimensional (3D) cell cultures, which involve embedding cells in a gel or another 3D material (see Justice, Badr, & Felder, 2009). Although these methods offer environmental conditions for cells that are close to those of intermediate tissues in vivo, the various physical parameters of the cell environment (geometry adhesion, environment elasticity, etc.) are not homogeneous and well defined, making it difficult to decipher their respective roles. Moreover, observation is challenging since cells can migrate in the three dimensions, limiting the possibilities of high-resolution microscopy. To circumvent these limitations, other methods have been developed. Among them, many have been driven to a large extent by technological developments that allow controlling forces or geometries at the micrometer scale, for example, apparatuses such as atomic force microscopes, microplates, or micromanipulators that have been designed by physicists to manipulate single cells and constrain them in various manners (see Mitrossilis et al., 2010). The main advantage of these approaches is the possibility to dynamically measure and control forces on a single cell. On the other hand, statistical and molecular analysis of a single cell is challenging. Microfluidic channels offer a different strategy to answer specific questions using simple ways to confine cells in a given geometry (see Velve-Casquillas et al., 2010). These channels allow the observation of many cells at the same time and offer tools to control specifically the microenvironment of the cell (chemical gradients, temperature, electric fields, etc.). Even though these microchannels are suitable to obtain statistics and build complex environments, the dynamic change of cell confinement and molecular analyses remain challenging in these conditions.

Here we describe an intermediary method that allows confining a population of adherent cells in a culture without the need to detach them from their original culture substrate. Because the confinement is applied on a population of cells, statistical data are easy to obtain, and because confinement can be released, the cell material is usable for further study (immunofluorescence, Western blot, qPCR [quantitative polymerase chain reaction], or other molecular analysis). The method is based on the gentle application of a confining slide on the cultured cell (Fig. 14.1A) using a

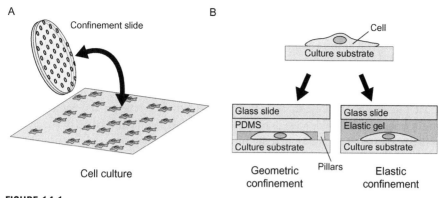

FIGURE 14.1

Principle of dynamic confinement with confinement slides. (A) Confinement slides are glass coverslips modified to provide a homogeneous confinement. They can be applied reversibly on the cell culture. (B) Slides can be modified either with a microstructured layer or PDMS, providing a geometrical confinement, or with a soft gel, providing an elastic confinement. (See color plate.)

suction cup device or a modified multiwell plate. The confinement slides are made of a glass coverslip, which is covered by a layer of material ensuring confinement. We will show how to use the microstructured layer of polydimethylsiloxane (PDMS), which is a relatively stiff biocompatible silicone rubber, to confine cells between two parallel planes (Fig. 14.1B). In this case, cells can move in two dimensions and the height of the cells is precisely imposed (geometric confinement). Alternatively, we will also show how to use a layer of polyacrylamide (PAM) gel to impose on the cell an environment of controlled stiffness (elastic confinement). In this second case, cells are covered by a gel of controlled rigidity and have to push on the gel to define their own space. In both cases, the rigid glass slide provides homogeneous confinement with no long-range deformation and the soft material (PDMS rubber or gel) ensures the robustness of the confinement by absorbing the local substrate inhomogeneities due to the possible presence of dust or particles. Except microfabricated molds, which need photolithography equipment, the fabrication of the confinement slides and handling devices is simple enough to be done in a cell biology lab with a small investment. Thus, this confinement method is well adapted to study of the roles of geometry and cell environment in cell processes such as migration (see Le Berre et al., 2013), division (see Lancaster et al., 2013), or adaptation (see Le Berre, Aubertin, & Piel, 2012).

The first section describes how to design microfabricated molds to produce confinement slides for geometric confinement. The following two sections specify the fabrication of the slides (for geometric confinement or gel confinement, respectively). Finally, two sections explain methods to handle these slides (either dynamically with a suction cup device or in multiwell plates to assay multiple conditions at the same time) followed by a discussion of how these slides can provide more complex configurations of confinement.

14.1 SLIDE DESIGN FOR GEOMETRICAL CONFINEMENT

For geometrical confinement, microspacers have to be molded under the confinement slides in a PDMS layer in order to control the distance between the slide and the culture substrate.

To get a robust confinement, the geometry of these spacers needs to be controlled. If the pillars are too close, many cells will be squashed by the pillars, but if they are too sparse, the glass slide sags in between the pillars, and the thickness of the cells is weakly controlled. In addition, if the surface of the pillars is large, a large proportion of the cells will be squashed, but, if the pillars are too small, they deform under pressure, giving an uncontrolled confinement height.

An optimal geometry of these spacers has been determined by *in silico* mechanical simulations to get pillars as far as possible from each other while making negligible long-range deformation of the confinement geometry (<100 nm when a 10-kPa pressure is applied on the slide) and a small amount of squashed cells (6%) (see Le Berre et al., 2012). Typically, spacers of 440 µm in diameter spaced by 1 mm in a hexagonal network provide meet these specifications (see Fig. 14.2A). Note that this geometry is adapted to confine cells to thicknesses above 3 µm. Below this height, sagging is more pronounced, and a capillary effect often sticks the PDMS layer on the culture substrate, squashing the cells on the whole surface.

Based on this geometry, we need a mold to imprint the structures on the confinement slides. This mold can be produced by photolithography of a layer of photoresist deposited on a silicon wafer by using standard microfabrication protocols. These protocols require a photomask that is used to shade a layer of photoresist from

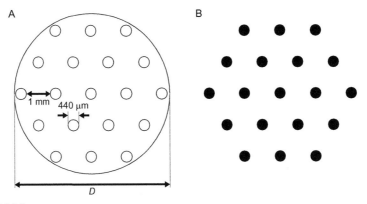

FIGURE 14.2

Optimal geometry of the pillars for geometrical confinement. (A) Dimensions of the pillars in a round slide of diameter *D*. Typically, *D*=10 mm for the suction cup device, and *D*=18 mm for a 6-well plate, and *D*=12 mm for a 12-well plate. The number of pillars has to be adapted depending on *D*. (B) Photomask to produce a negative photoresist layer (such as SU8) corresponding to the design in (A).

the applied UV light and define the shape of the spacer after development of the photoresist. With this fabrication mode, the thickness of the photoresist layer corresponds to the height of the pillars and, therefore, the thickness of the cells after confinement. If manipulated carefully, a mold can be used many times. Thus, it is not necessary to have access to a microfabrication facility, and the mold, which is technically easy to fabricate, can be created by a specialized collaborator or a microfluidic foundry company. The mold provider usually needs a drawing of the photomask in DXF (Drawing eXchange Format) or GDSII (Graphic Database System II) format, which can be done easily with CAD software like AutoCAD (Autodesk) or L-Edit (Tanner). But due to the poor resolution required to produce the mold, the mask can also be printed on a plastic sheet with a high-resolution printer by using a simple drawing software like Adobe Illustrator. This way, many academic microfabrication facilities can make such a mold for less than $200. (For more details on mold fabrication, see, for example, Velve-Casquillas et al., 2010.)

14.2 GEOMETRICAL CONFINEMENT SLIDE FABRICATION

This section describes the fabrication of stiff confinement slides for geometrical confinement.

14.2.1 Materials

- A mold to imprint slides (see previous section 14.1).
- 10-mm-diameter standard glass coverslips.
- PDMS (RTV615, General Electrics or Sylgard 184).
- Wash bottle of isopropanol.
- Scalpel blade.

For slide treatment (optional):

- Parafilm.
- Wash bottle of pure water.

Adhesive:

- Fibronectin solution 50 µg/ml in phosphate buffered saline (PBS) (30 µl per slide) (Fibronectin from Sigma).

Nonadhesive:

- Pll-g-PEG solution 500 µg/ml in HEPES (4-(2-hydroxyethyl)-1-piperazineethanesulfonic acid) buffer 10 mM, pH 8.6 (30 µl per slide) (pLL-g-PEG (20)-[3.5]-(2) from SuSoS).

For antiadhesive coating of the mold (optional):

- Trimethylchlorosilane (TMCS) (Sigma).
- Fume hood.

14.2.2 **Equipment**

- Plasma cleaner (for example, PDC-32G, Harrick).
- Pressurized, filtered, and oil-less air stream.
- Hot plate.

14.2.3 **Fabrication (Fig. 14.3A–F)**

1. Prepare 10 g of PDMS mix with a ratio of PDMS/cross-linker (A/B) of 8/1 (w/w), avoiding introducing too many bubbles during mixing. PDMS can be mixed either manually with a simple swizzle stick or with an automatic planetary mixer (e.g., ARE-250, Thinky). If the PDMS is mixed manually, bubbles can be removed by centrifugation (2 min at 1000 rpm). Be careful in

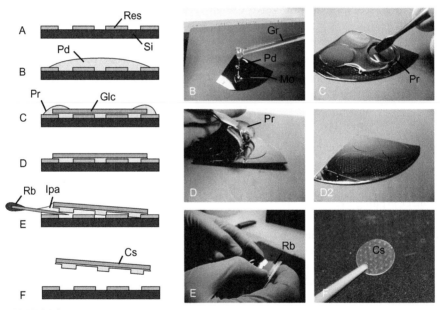

FIGURE 14.3

Confinement slide fabrication. The schemes and corresponding photos summarize the fabrication process: (A) Start with a microfabricated mold (Mo) made of a silicon wafer (Si) covered by a patterned layer of photoresist (Res). (B) A drop of PDMS (Pd) is placed on the mold, for instance, with the help of a glass rod (Gr). (C) A plasma-activated glass coverslip (Glc) is pressed onto the drop of PDMS with the help of tweezers in order to keep only a residual layer of PDMS under the coverslip. The excess PDMS should make rims around the coverslips (Pr). (D) After baking, the PDMS rims (Pr) are removed. (D2) No residual PDMS should be visible on the mold at this stage. (E) To unmold the confinement slide (Cs), the slide is wet in isopropanol (Ipa) before a razor blade (Rb) is introduced between the mold and the slide. (F) After unmolding, the confinement slide (Cs) is rinsed with isopropanol and air dried. (See color plate.)

this case that PDMS is well mixed because unreticulated PDMS is very difficult to remove and can make your mold unusable.

2. Place the 10-mm glass coverslips on a big glass slide and treat it in the plasma cleaner for 2 min at maximum power in order to activate one side of the slide (when the plasma is on, the plasma may produce a dense pink glow). This step allows the glass slide to stick onto the PDMS layer in the next step.

3. With a glass rod or equivalent, distribute drops of PDMS on the mold in the places you want to put the plasma-treated coverslips (Fig. 14.3B).

4. Place the coverslips on top of the drop and push gently with tweezers everywhere on top of the coverslips to keep only a very thin layer of PDMS under the coverslip. If you push too hard on the coverslip, it will break, and, if you push too gently, the residual layer of PDMS will be too thick, and the pillars will collapse. To gauge the right pressure, notice that it corresponds to the minimum pressure where you see a small white star appearing under the tweezers. Do not use sharp tweezers: You will break your coverslips! The ejected PDMS should form a rim partially covering the slide (Fig. 14.3C).

5. Bake the mold with the coverslips on the hot plate at 95 °C for 15 min. At this step, if the PDMS is baked too long or not long enough, it will stick too strongly to the mold, and it will be very difficult to unstick confining slides without breaking them.

6. Remove the PDMS rim with tweezers (Fig. 14.3D).

7. Place a droplet of isopropanol on the mold; it will help to unmold the slide.

8. Use a razor blade to gently unstick the coverslip from the mold. To avoid scratching the mold (molds made of photoresist are very easy to scratch) or breaking the coverslip, use the following method: At the beginning, place the razor blade on the edge between the mold and the coverslip at an angle of 20°, and then carefully slide the blade under the confining coverslip parallel to the mold surface (Fig. 14.3E).

9. Rinse the coverslip with isopropanol.

10. Dry the confining coverslip with an air gun (otherwise it will leave stains and inhomogeneities on the confining slide surface) (Fig. 14.3 F).

11. Gently clean the mold with a clean wipe soaked with isopropanol.

12. *Optional*: If the mold becomes too sticky and coverslips are difficult to detach, use an antiadhesive agent. For this step, work under a fume hood: TMCS is very toxic. Carefully clean the mold with isopropanol and a soft cleaning paper wipe (e.g., Kimtech, Kimberly-Clark), and put it in a petri dish. Seed two or three drops of TMCS around the mold, and close the petri dish. The drops will evaporate and react with the surface. After 5 min, open the petri dish, and vent the remaining TMCS vapors for 5 min, then bake the mold for 10 min at 70 °C to stabilize the surface.

14.2.3.1 Slide treatment

The coverslip can be used either without any treatment (it is not toxic to the cells) or with a chemical treatment to promote or avoid cell adhesion. However, we advise treating the confining slide systematically in order to avoid trapping air bubbles between the slide and the substrate, which can kill the cells (PDMS is very hydrophobic and often retains air bubbles on its surface).

1. For nonadhesive treatment only: Place the confinement slide on a big glass slide with the PDMS side facing upward, and treat it in the plasma cleaner for 30 s maximum at maximum power in order to activate the PDMS surface.
2. On a piece of parafilm, drop 30 μl of the fibronectin solution (for adhesive treatment) or of the pLL-g-PEG solution (for nonadhesive treatment).
3. Place the confinement slide on the drops, the structured slide in contact with the solution, and wait for 30 min.
4. Rinse extensively with pure water.
5. Dry the slide with an air jet.

14.3 SOFT CONFINEMENT SLIDE FABRICATION

This section describes the fabrication of soft confinement slides for elastic confinement.

14.3.1 Materials

a. Coverslips (any glass coverslip size can be used).
b. Ethanol 70%.
c. Deionized water.
d. Acrylamide 40% (Biorad).
e. Bis-acrylamide 2% (Biorad).
f. 10 mM HEPES buffer pH 7.6.
g. Irgacure 0.5% in 10 mM HEPES pH 7.6 (Irgacure 2959, CIBA). The Irgacure stock solution has to be protected from light.
h. Aminopropyltrimethoxysilane (APTMS) (Sigma) 4% solution in acetone.
i. Clean and smooth PMMA (polymethyl methacrylate) plate (Good Fellow) (Also available under the commercial name of "plexiglas").
j. PDMS (RTV615, GE).
k. Wash bottle of isopropanol.

14.3.2 Equipment

l. Ultrasonic bath.
m. Plasma cleaner (e.g., PDC-32G, Harrick).
n. UV lamp (e.g., Delolux 03 S).
o. Fume hood.
p. Vacuum bell cloche.

14.3.3 **Method**

1. *Coverslip surface preparation*
    a. Wash the glass coverslip with ethanol 70% and dry it.
    b. Expose the dry coverslips to air plasma for 30 s, maximum power.
    c. Incubate clean slides in a solution of APTMS (4% in acetone) for 30 min under sonication.
    d. Wash with distilled water, and air dry (slides can be stored for several weeks at room temperature).
    e. Place the slide on a drop of glutaraldehyde solution (0.5% in PBS) for 30 min. It is convenient to do it on a laboratory parafilm. Drops of 100 μl are sufficient for coverslips of 18 mm. Keep track of the side of the coverslip that was treated. Coverslips should be used right after this step. (Glutaraldehyde is toxic and an irritant, and it has to be manipulated under a fume hood.)
    f. Wash coverslips with distilled water, and air dry.
2. *Gel polymerization (Fig. 14.4A–E)*
    a. Prepare the PAM mix in a stock solution of Irgacure (0.5% in 10 mM HEPES pH 7.6). The concentration of acrylamide and bis-acrylamide depends on the gel rigidity you want to obtain (see Table 14.1 and Boudou, Ohayon, Picart, & Tracqui, 2006).

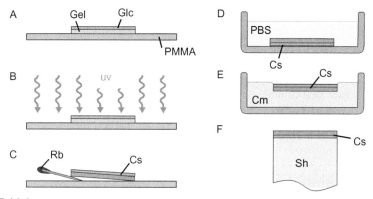

FIGURE 14.4

Fabrication of confinement slides with gels of controlled rigidities. (A) First a droplet of the gel precursor mix (Gel) is deposited on a PMMA plate and covered by a pretreated glass coverslip (Glc). (B) The gel is polymerized under UV. (C) After polymerization, the confinement slide (Cs) can be detached from the PMMA plate with a razor blade (Rb). (D) The gel is incubated in PBS to remove any remaining diffusible molecules. (E) Before the experiment, the gel is incubated in the appropriate culture medium (Cm) upside down on the surface of the medium to avoid getting the back side wet in the medium. (F) When positioned on the confinement apparatus, the dry side of the slide sticks naturally to the PDMS surface of the slide holder (Sh). (See color plate.)

Table 14.1 Quantities of acrylamide, bis-acrylamide, and irgacure solution used to get 1 ml of gel depending on the required rigidity

Young's modulus (kPa)	Irgacure 0.5% (μl)	Acrylamide 40% (μl)	Bis-acrylamide 2% (μl)
2	840.0	125.0 (5%)	35.0 (0.07%)
5	784.5	185.5 (7.5%)	30.0 (0.06%)
10	767.0	185.5 (7.5%)	47.5 (0.95%)
15	749.5	185.5 (7.5%)	65.0 (0.13%)
20	714.5	185.5 (7.5%)	100.0 (0.2%)
30	650.0	250.0 (10%)	100.0 (0.2%)

b. Put drops of the PAM solution on the PMMA plate, and gently place the coverslips on them with the treated surface toward the PAM solution (the coverslip should float on the drop). The volume of the drops can be adapted according to the thickness of the gel you want to obtain (200 μl for an 18-mm coverslip will result in gels of ~0.8 mm during the fabrication, but note that the gel will swell to ~2 mm after incubation) (Fig. 14.4A).

c. Place the setup under a UV lamp for 10 min (Fig. 14.4B).

d. Carefully detach the coverslips with gel from the plastic using a razor blade or a scalpel (Fig. 14.4C), and quickly place it in sterile PBS for incubation overnight (Fig. 14.4D).

e. Before experiments with cells, gels have to be incubated at least 2 h in the cell culture medium that will be used during the experiment. If drugs will be used in the experiment, the gel has to be incubated in the medium containing the appropriate concentration of the given drug. To avoid wetting the back side of the confinement slide, it can be deposited upside down on the medium surface. This way, the slide will be incubated floating on the surface (Fig. 14.4E).

14.4 DYNAMIC CELL CONFINER

The application of the confinement slide on the cultured cells can be done in several ways. In all cases, a slight pressure (~10 kPa) has to be applied on the slide to get a robust and uniform contact with the substrate of the cultured cells, and tangential movement must be eliminated to avoid shearing the cells. A practical way to proceed is to use the so-called cell confiner that is described in this section.

The cell confiner is a device that acts as a suction cup, as shown in Figure 14.5. It is composed of a soft empty part that forms a sealed cavity when it is placed on a surface. When a vacuum is created in the sealed cavity, the device is stuck on the substrate and cannot move. In the center of the empty part, a piston holds the confinement slide, which can be displaced vertically by changing the strength of the vacuum in the device. After the threshold strength of the vacuum is applied, the confinement slide comes into contact with the cultured cells.

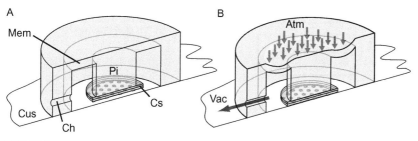

FIGURE 14.5

Principle of a dynamic cell confiner. (A) The cell confiner acts as a suction cup, which can be stuck onto a cell culture substrate (Cus). The pressure in the device is controlled through a connection hole (Ch). When a vacuum is applied in the device (Vac), the atmospheric pressure (Atm) maintains the device on the culture substrate. A piston (Pi) in the central part of the confiner holds the confinement slide (Cs) face to the culture substrate. The position of this piston relative to the substrate can be precisely tuned by controlling the strength of the vacuum in the device, thanks to the deformation of the membrane (Mem) holding the piston. (B) The confinement slide can be pressed gently onto the cell population by aspirating air through the connection hole. (For a color version of this figure, the reader is referred to the online version of this chapter.)

By using this device, confinement slides can be applied onto any substrate provided that it is flat. The culture substrate just has to be large enough to accept the device (the schematic shown in Fig. 14.6A is adapted to 35-mm-diameter petri dishes).

14.4.1 **Materials**

For fabrication:

- Machined confiner mold: This mold, intended to mold the PDMS device, can be fabricated in stainless steel in a standard workshop. In the simplest version, the mold is composed of two stainless rings, a silicon wafer (which can be replaced by a glass slide), and a 0.5-mm-thick, round glass coverslip of 16 mm diameter (which can be created by gluing several thinner glass coverslips together), according to the layout shown in Figure 14.6B.
- PDMS (RTV615, GE).
- A standard glass slide (typically 1×3 in., 1 mm thick).
- A hot plate.
- A 0.75-mm puncher to make the vacuum inlet hole in the device (Elveflow).

For handling:

- The cell confiner.
- A confinement slide.
- 70% alcohol.
- Absorbing paper.

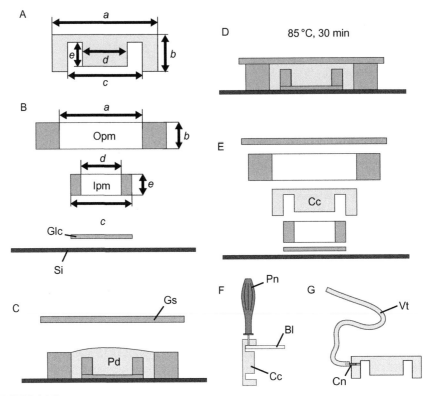

FIGURE 14.6

Cell confiner fabrication. (A) Dimensions of the cell confiner. Typically, for a cell confiner adapted to 35-mm dishes: $a=22$ mm; $b=7$ mm; $c=16$ mm; $d=8$ mm; $e=5$ mm. (A) A mold can be easily fabricated by assembling on a silicon wafer (Si) a 0.5-mm glass coverslip (Glc) and two rings forming the outer part of the mold (Opm) and the inner part of the mold (Ipm). (B) After the mold has been filled with PDMS (Pd), a glass slide (Gs) is pressed on the top of the structure to flatten the device surface and produce its final shape. (C) The device is baked for 30 min at 85°. (D) After cooling down, the structure can be disassembled to retrieve the molded cell confiner (Cc). (F) A 0.75-mm hole is punched into the wall of the device with a puncher (Pn) by inserting a blade (Bl) below the drilling zone for protection. (G) This hole is used to insert a fluid connector (Cn) plugged into the vacuum tube (Vt). (For a color version of this figure, the reader is referred to the online version of this chapter.)

- Removable adhesive tape (e.g., Scotch Magic Tape, 3 M).
- A precision vacuum generator or controller (e.g., Elveflow VG1006). Check before the experiment that you have all the connections plugged into the device. Sometimes, the culture medium is aspirated into the controller by the vacuum, which can cause severe damage to the controller. To avoid this, we strongly recommend placing a security reservoir between the device and the vacuum controller.

- Adapted tubing to plug into the vacuum generator, including a 0.51-mm ID tygon tube (Saint-Gobain) and Stainless Steel Tubing 0.025″ OD (see, e.g., microfluidic kits, Elveflow) as a connector to plug the tube into the device (see Fig. 14.6G). It is best to bend the tube to avoid it taking up too much space in the petri dish.
- A sharp pair of tweezers.

14.4.2 Design and Fabrication

The design shown in Figure 14.6A is for a 35-mm petri dish. This particular geometry has been optimized to get a working pressure of −10 kPa (−100 mbar), which is sufficient to handle the device easily, but does not notably change the pressure around the cells. For other formats of cell cultures, the device can be scaled up or down: The working pressure of the device is independent of the device size if the proportions are respected.

The suction cup device can be molded with PDMS in a simple mold provided the top and bottom surfaces of the mold are flat enough to get a good optical transparency. Here, we describe a simple method to make a mold based on the simplest pieces, which can be done in any workshop.

1. Prepare a mix of PDMS/cross-linker (10/1 w/w) with no bubbles. (If you do not have a planetary mixer, you can remove bubbles either with a vacuum bell cloche for 2 h or by centrifuging PDMS for 2 min at 1000 rpm.)
2. Stack the different parts of the mold directly on a (cold) hot plate (Fig. 14.6B).
3. Pour PDMS very slowly into the mold, avoiding bubble formation (Fig. 14.6C).
4. Cover the filled mold with a glass slide, and press on it until there remains only a residual layer of PDMS between the mold and the slide (Fig. 14.6D).
5. Heat up the hot plate to 85 °C for 30 min (Fig. 14.6D).
6. Cool down the molded part before unmolding. To make unmolding easier, use a razor blade and isopropanol to reduce adhesion of the PDMS on the mold.
7. Punch the inlet hole with the 0.75-mm puncher as shown in Figure 14.6 F.

14.4.3 Handling

1. At least 1 h before the experiment, incubate the confinement slide in the cell culture medium that will be used for the experiment to equilibrate the PDMS with the medium. The structured side (PDMS side) of the slide should face upward. Be careful not to lose the orientation of the slide during manipulation; it can be difficult to recover. Note that PDMS absorbs small hydrophobic molecules from the medium. So, if drugs are used in the experiment, these drugs have to be present during this incubation step.
2. Before the experiment, clean the cell confiner with 70% alcohol, and dry it carefully with absorbing paper. Be careful not to leave the device in contact with the alcohol for a long time, or it will absorb the alcohol and become toxic to cells.
3. Clean the bottom face of the device with removable adhesive tape to remove dust particles.

4. Plug the device into the controllable vacuum source, and tune the controller to −3 kPa (−30 mbar). At this pressure, and with a good seal, the device sticks to the substrate, but the confinement slide does not yet touch the cells.
5. Prepare an absorbing paper to dry the back side of the confinement slide.
6. Pick up the confinement slide with tweezers, and dry its back side on the absorbing paper by placing it onto the paper. (If the back side is not totally dry, it will not stick onto the device and can be lost during handling.)
7. Place the confinement slide onto the piston of the cell confiner, structured side up. The slide should stick naturally onto the piston.
8. Put the device in contact with the cell culture substrate. Take care not to deform the device to achieve a good seal of the device cavity. Also, be careful not to push on the center of the device, or cells could be crushed by the device. After the device is well sealed, the membrane sustaining the piston should be slightly deformed. The step of placing the device must be carried out quickly, otherwise the medium will fill the device and enter the vacuum outlet. If the medium enters the vacuum tube, the device may not work properly due to the capillary pressure of the meniscus in the tube. If the medium has entered the tube by accident, the device and the tube must be cleaned and dried to avoid the formation of a meniscus in the tube.
9. At this point, cells can be observed before confinement. The petri dish will not move during the confinement operation, and several cells can be observed simultaneously.
10. To confine cells, slowly decrease the pressure down to −10 kPa (−100 mbar). By observing the cells during this time, one should see the confinement slide touching cells at approximately −5 kPa.
11. Confinement can be released if the pressure is increased back to −3 kPa.

14.5 MULTIWELL CONFINER

If a large amount of confined cells or multiple conditions are required, a modified multiwell plate can be used to apply confinement slides on cells (Fig. 14.7). In this case, it is not the force induced by a vacuum that holds the slide on the cells, but the pressure induced by the deformation of a soft pillar. In this case, big PDMS pillars are stuck on the lid of the multiwell plate and hold the confinement slides. When the lid is closed, the pillars push the confinement slides onto the culture substrate and confine the cells. Many wells can be processed simultaneously, and confinement slides of larger surfaces can be used; however, the speed of confinement is not controlled, and it is not convenient to observe what happens during the confinement application. For this reason, this method is not applicable to the study of short-time events. In contrast, it is well adapted to long-term experiments and molecular analysis (e.g., overnight time-lapse imaging, immunofluorescence).

14.5.1 Materials

For fabrication:
- A 12 (or 6) well multiwell plate (TPP).
- A 48 (or 12) well multiwell plate.

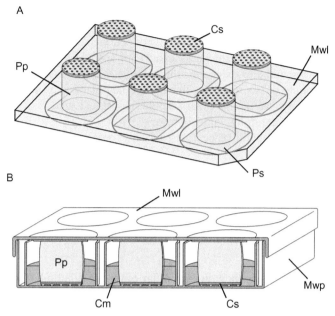

FIGURE 14.7

Multiwell plate confiner. (A) A lid of a multiwell plate (Mwl) is modified by introducing large PDMS pillars (Pp) to hold the confinement slides (Cs). These pillars typically have the height of the plate, which can be slightly increased with a PDMS sheet (Ps). (B) When the multiwell plate (Mwp) containing cells and culture medium (Cm) is closed with the modified lid, the confinement slides are applied on the cells adhering to the bottom of the wells. Pillars that are slightly higher than the well depth deform and apply pressure on the confinement slides. (For a color version of this figure, the reader is referred to the online version of this chapter.)

- PDMS (RTV 615, GE).
- Wash bottle of isopropanol.
- Removable adhesive tape (Scotch Magic Tape, 3 M).
- One sheet of gel-pack film (DGL Film X8 17 mils, Gel-Pack).

For handling:
- Adhesive tape.

14.5.2 Fabrication

1. Prepare the PDMS mix with a ratio of 35:1 w:w PDMS:cross-linker. Due to the small concentration of cross-linker, the PDMS has to be mixed extensively to get a good reticulation.
2. Pour the PDMS mix into the mold, which can be either a smaller multiwell plate or a custom-made mold. For example, for pillars to be used in 12-well plates, use

a 24-well plate as mold and fill them with PDMS. Pour the PDMS in excess, leave it in the vacuum bell cloche to get rid of bubbles, and carefully close with the lid of the plate avoiding bubble creation. For confinement slides including PAM gels, which have a higher thickness, spacers of appropriate thickness can be stuck on the plate lid to reduce the thickness of the final pillars.

3. A small amount of isopropanol will help get the pillars out of the mold. After unmolding and cleaning, the pillars should be soft and sticky.
4. Clean the pillars carefully with isopropanol.
5. On the multiwell plate lid, place a gel-pack sheet on each well position in order to raise the pillars slightly.
6. Place the pillars on the gel-pack sheet at the center of each well position. Clean pillars will stick naturally. You can clean the PDMS and gel-pack sheet before use with the adhesive tape to remove dust.
7. Now the lid is ready to be used.

14.5.3 Handling

1. Place one confinement slide on each pillar of the modified lid.
2. Optional: sterilize the lid under UV (365 nm).
3. Optional: incubate the pillars in culture medium to equilibrate the PDMS.
4. Gently close the multiwell plate with the modified lid.
5. While pushing on the lid, fix the lid to the multiwell plate with adhesive tape. During this step, avoid moving the lid laterally: This will shear cells and kill them. Pull on the adhesive tape while applying it to the closed plate to put the plate under tension.
6. Verify under a microscope that the cells are well confined.

14.6 DISCUSSION

In this chapter, we have discussed how to confine cells in between two parallel surfaces separated by a defined gap in a reversible manner. Alternatively, we explained an adaptation of the method to confine cells below a soft PAM gel of controlled stiffness. However, since the two surfaces can be treated independently, and because the treatment of the confinement slides can be very versatile, this method is not limited to these two configurations in its principles; an infinity of confinement configuration can be imagined. The method can be easily combined with others to build more complex environments for specific studies. For the geometric confinement, for example, the flat roof can be replaced by more complex geometric features by using advanced photolithography methods (see del Campo & Greiner, 2007). Thus, it might be possible to introduce various textures or restrictions into the geometric landscape of the cell environment. Another parameter that can be varied is the level of adhesion. This can be done by mixing in a radiometric manner pLL-g-PEG with pLL-g-PEG-RGD, which is a modified version of pLL-g-PEG modified with the RGD motif and is responsible for integrin-mediated adhesion (see Barnhart, Lee, Keren, Mogilner, & Theriot, 2011). In this case, the adhesion level of cells can be modulated independently on both sides of the cell. Confinement can also be combined with other methods. For example, it is possible to print adhesive micropatterns on the bottoms of the wells to restrict geometric cell adhesion. This allows a more reproducible organization of the cell under confinement (see Azioune, Carpi, Tseng, Théry, & Piel, 2010).

GENERAL CONCLUSION

Confinement parameters are taking on increasing importance in cell biology studies, and tools that allow the control of the different confinement parameters independently while remaining simple enough to be used in a biological routine are crucial. We described in this chapter a versatile way to confine a population of cells in a controlled environment (geometry, stiffness, adhesion, etc.), which is easily combinable with specific observation techniques and molecular biology protocols. We hope that these protocols will help the community answer certain questions and will lead to new and innovative ideas about how to study cells in complex environments.

References

Azioune, A., Carpi, N., Tseng, Q., Théry, M., & Piel, M. (2010). Protein micropatterns: A direct printing protocol using deep UVs. *Methods in Cell Biology, 97*, 133–146.

Barnhart, E. L., Lee, K. C., Keren, K., Mogilner, A., & Theriot, J. A. (2011). An adhesion-dependent switch between mechanisms that determine motile cell shape. *PLoS Biology, 9*(5), e1001059.

Boudou, T., Ohayon, J., Picart, C., & Tracqui, P. (2006). An extended relationship for the characterization of Young's modulus and Poisson's ratio of tunable polyacrylamide gels. *Biorheology, 43*, 721–728.

del Campo, A., & Greiner, C. (2007). SU-8: A photoresist for high-aspect-ratio and 3D submicron lithography. *Journal of Micromechanical Microengineering, 17*, R81–R95.

Justice, B. A., Badr, N. A., & Felder, R. A. (2009). 3D cell culture opens new dimensions in cell-based assays. *Drug Discovery Today, 14*(1–2), 102–107.

Lancaster, O. M., Le Berre, M., Dimitracopoulos, A., Bonazzi, D., Zlotek-Zlotkiewicz, E., Picone, R., et al. (2013). Mitotic rounding alters cell geometry to ensure efficient bipolar spindle formation. *Developmental Cell, 25*(3), 270–283.

Le Berre, M., Aubertin, J., & Piel, M. (2012). Fine control of nuclear confinement identifies a threshold deformation leading to lamina rupture and induction of specific genes. *Integrative Biology, 4*(11), 1406–1414.

Le Berre, M., Liu, Y.-J., Hu, J., Maiuri, P., Benichou, O., Voituriez, R., et al. (2013). *Geometric friction directs cell migration,* unpublished.

Mitrossilis, D., Fouchard, J., Pereira, D., Postic, F., Richert, A., Saint-Jean, M., et al. (2010). Real-time single-cell response to stiffness. *Proceedings of the National Academy of Sciences of the United States of America, 107*(38), 16518–16523.

Velve-Casquillas, G., Le Berre, M., Piel, M., & Tran, P. T. (2010). Microfluidic tools for cell biological research. *Nano Today, 5*(1), 28–47.

Benzophenone-Based Photochemical Micropatterning of Biomolecules to Create Model Substrates and Instructive Biomaterials

15

Aurora J. Turgeon*, Brendan A. Harley[†], and Ryan C. Bailey*

*Departments of Chemistry, University of Illinois at Urbana-Champaign, Urbana, Illinois, USA
†Chemical and Biomolecular Engineering, University of Illinois at Urbana-Champaign, Urbana, Illinois, USA

CHAPTER OUTLINE

Methods in Cell Biology, Volume 121
ISSN 0091-679X
http://dx.doi.org/10.1016/B978-0-12-800281-0.00015-4

Abstract

The extracellular matrix (ECM) is a dynamic and heterogeneous environment that controls many aspects of cell behavior. Not surprisingly, many different approaches have focused on creating model substrates that recapitulate the biomolecular, topographical, and mechanical properties of the ECM for *in vitro* studies of cell behavior. This chapter details a general, versatile method for the spatially controlled deposition of multiple biomolecules onto both planar and topographically complex support structures with micrometer resolution. This approach is based upon the well-understood photochemical UV crosslinking of benzophenone (BP) to solution-phase biomolecules. This is a molecularly general strategy that can be utilized to immobilize biomolecules onto any surface prefunctionalized with BP. Examples described herein include modification of planar and corrugated glass substrates as well as collagen–glycosaminoglycan biomaterials configured either as highly porous scaffolds or nonporous membranes with a variety of biomolecular targets, including proteins, glycoproteins, and carbohydrates.

INTRODUCTION

Every living cell resides in an enormously complex and dynamic environment responsible for biological and biophysical cues that govern fate and function. This cellular niche is comprised of matrix proteins, proteoglycans, and other biomolecules working in conjunction with mechanical stimuli and paracrine signals to dictate a myriad of cell behavior such as adhesion, inflammatory response, and stem cell differentiation (Tibbitt & Anseth, 2012). The ability to mimic this complex cellular niche is of broad interest within the biomaterials research community.

We have developed a simple approach that takes advantage of the photochemical properties of benzophenone (BP). Upon illumination with 365-nm light, BP undergoes an $n \rightarrow \pi^*$ transition to form a transient diradical that can covalently attach to any proximal solution phase biomolecules to the surface via insertion into a C–H bond. When BP is immobilized to the surface of a substrate, the new C–C bond represents a covalent tether between the biomolecule and the substrate. If BP does not react with a nearby molecule within the excited state lifetime (up to $\sim$120 μs) it relaxes back to the ground state from which it can be re-excited with subsequent optical excitation. Capitalizing on the fact that attachment is driven by UV excitation, geometric patterns and gradients of biomolecules can be generated by modulating the location and duration of exposure incident onto a BP-modified substrate. This approach has been previously utilized by our group to generate single and multicomponent biomolecular patterns and gradients that were used for biological studies of cell–material interactions (Herman, Potts, Michael, Tolan, & Bailey, 2011; Martin, Caliari, Williford, Harley, & Bailey, 2011; Martin, Herman, et al., 2011; Toh, Fraterman, Walker, & Bailey, 2009). The general BP immobilization and UV patterning procedures are easily adaptable for the patterning of almost any biomolecular target and convenient conjugation chemistry makes it broadly amenable to a range of substrate materials. One particularly interesting adaptation is the extension from patterning on two-dimensional substrates to three-dimensional materials, which is uniquely enabled by this approach because physical contact is not required for patterning, as is the case for microcontact-based biomolecular attachment strategies. Photopatterning on corrugated glass substrates and three-dimensional porous scaffolds further demonstrate the generality of the approach and open up potential avenues to studying the interplay of biomolecular cues, surface texture, and mechanical properties.

15.1 BP PATTERNING ON GLASS SLIDES

This process is based upon previously described methods (Herman et al., 2011; Martin, Herman, et al., 2011; Toh et al., 2009). It has been successfully applied to the spatial patterning of a diverse set of proteins, glycoproteins, and carbohydrates, including concanavalin A, P- and E-selectin, mannan, fibronectin, and ICAM-1. A schematic illustration of the surface chemistry and photopatterning scheme, along with images showing representative biomolecular patterns is shown in Fig. 15.1.

In addition to patterning on planar microscope slides, corrugated glass substrates, created by photolithography and etching, may be substituted (Martin, Herman, et al., 2011). This section presents two approaches to surface chemical modification that omits or includes poly(ethyleneglycol) (PEG) moieties, as PEG is known to reduce nonspecific binding of cells and proteins to surfaces (Harder, Grunze, Dahint, Whitesides, & Laibinis, 1998; Raghavan, Desai, Kwon, Mrksich, & Chen, 2010).

15.1.1 Materials

- Glass microscope slides (Fisher Scientific, Philadelphia, PA)
- Piranha solution (4:1 (v:v) concentrated H_2SO_4: 30% H_2O_2)
- Absolute ethanol (Decon Laboratories, King of Prussia, PA)
- 3-(Triethoxysilyl)butyl aldehyde (Gelest, Morrisville, PA)

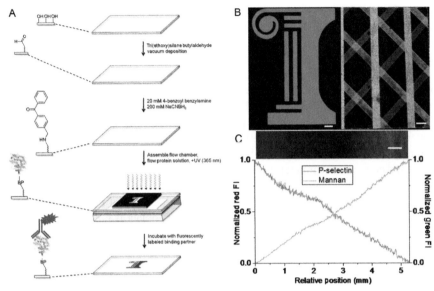

FIGURE 15.1

Photoimmobilization of biomolecules onto glass substrates. (A) Schematic illustration showing the surface chemistry and subsequent approach to photopatterning. (B, left) Photoimmobilized "Illinois logo" pattern of biotinylated concanavalin A visualized with fluorescently labeled streptavidin. (B, right) Three-component pattern of mannan (stripes running from top right to bottom left), P-selectin (stripes running from top left to bottom right), and fibronectin (vertical stripes). Scale bars: 100 μm. (C) A single substrate displaying two overlapping gradients of P-selectin (high left, low right; red) and mannan (low left, high right; green). FI stands for fluorescence intensity. Scale bars: 500 μm. (See color plate.)

Reprinted with permission from Martin, Herman, et al. (2011). Copyright 2011 American Chemical Society and Toh et al. (2009). Copyright 2009 American Chemical Society.

- 4-Benzoyl benzylamine hydrochloride (Matrix Scientific, Columbia, SC)
- Sodium cyanoborohydride (Sigma-Aldrich)
- Dimethylformamide (DMF) (Fisher Scientific)
- Methanol (Fisher Scientific)
- Aldehyde-blocking buffer (0.1 M Tris, 200 mM ethanolamine, pH 7.0)
- Biomolecules to be patterned
- Phosphate buffered saline (PBS) (Sigma-Aldrich)
- Bovine serum albumin (BSA) (Sigma-Aldrich)
- Tween 20 (Sigma-Aldrich)
- Glycine (Acros Organics)
- H_2N-PEG-CM (1000 MW, Laysan Bio, Inc., Arab, AL)
- N-(3-Dimethylaminopropyl)-N'-ethylcarbodiimide hydrochloride (EDC)
- N-Hydroxysuccinimide (NHS)
- NHS-quenching buffer (0.1 M Tris, 100 mM ethanolamine, pH 8.5)

15.1.2 **Equipment**

- Vacuum desiccator
- Goniometer (Ramê-Hart, Netcong, NJ)
- Hot plate
- Parallel flow chamber (GlycoTech; Gaithersburg, MD)
- Silicone gasket of 127 μm thickness, 6 cm length, and 1 cm width
- Syringe pump (Harvard Apparatus, Hollison, MA)
- UV light source (Preferred: Argon ion laser, Coherent Innova 90-4)
- Refractive beam shaping optics (π-Shaper, Molecular Technologies, Berlin, Germany)
- Chromium-coated quartz photo masks
- Opaque shutter
- Computer-controlled translational stage (Thorlabs Opto dc driver)
- Optical power meter

15.1.3 **Method**

15.1.3.1 *Surface functionalization: BP-modified glass slides*

1. Glass microscope slides are first cleaned with a Piranha solution (4:1 (v:v) concentrated H_2SO_4: 30% H_2O_2) (Lawrence & Springer, 1991). Heat solution and slides for 20 min.
2. Rinse substrates three times in water and once in ethanol by carefully dunking the slides in each rinse component. After rinsing, the slides are dried under a stream of nitrogen.
3. Bake slides in an oven at 120 °C for 1 h and then cool to room temperature.
4. The cooled slides are positioned upright along the wall of a vacuum desiccator. The cap of a microcentrifuge tube containing 100 μl of 3-(triethoxysilyl)butyl aldehyde is placed in the center and vacuum applied. The silane is allowed to deposit onto the glass slide for 2.5 h under vacuum.
5. Cure slides at 120 °C for 1 h.
6. Rinse in absolute ethanol for 30 min and dry under a stream of nitrogen.
7. Measure water contact angle measurements on a goniometer to confirm successful silanization. The contact angles should be approximately 45°.
8. Prepare a solution of 20 mM 4-benzoyl benzylamine hydrochloride and 200 mM $NaCNBH_3$ in 4:1 DMF:MeOH. Pipet 200 μl of this solution between a pair of glass slides such that the silanized slides of both slides are in contact with the inner liquid layer. Cover with foil and leave for 4 h.
9. Quench the reaction by separating the sandwiched slides and immersing into aldehyde-blocking buffer (0.1 M Tris, 200 mM ethanolamine, pH 7.0) for 1 h at room temperature.
10. Rinse thoroughly with water, DMF, methanol, and ethanol. Dry under a stream of nitrogen.

11. Measure contact angles to verify BP functionalization. The contact angles should be approximately 53°.
12. Store BP-functionalized slides in opaque slide holders in a desiccator at room temperature for up to a month.

15.1.3.2 Surface functionalization: BP-modified glass slides incorporating PEG

1. Treat slides as above through step 5.
2. Following silane curing, incubate slides for 4 h at room temperature in the presence of 10 mM H$_2$N-PEG-CM and 100 mM NaCNBH$_3$ in water using the slide sandwich technique described earlier.
3. Quench the reaction by immersion in aldehyde-blocking buffer for 1 h at room temperature.
4. Rinse slides with water, methanol, and ethanol. Dry under a stream of nitrogen.
5. Incubate slides in the dark for 4 h at room temperature in the presence of 20 mM 4-benzoyl benzylamine, 75 mM N-(3-dimethylaminopropyl)-N'-ethylcarbodiimide hydrochloride (EDC), 30 mM NHS in PBS (pH 7.4).
6. Immerse slides in NHS-quenching buffer (0.1 M Tris, 100 mM ethanolamine, pH 8.5) for 1 h at room temperature in the dark.
7. Rinse slides with water, methanol, and ethanol. Dry under a stream of nitrogen.
8. Store the resulting BP-modified substrates in a desiccator in the dark until further use.

15.1.3.3 Photoimmobilization of proteins or carbohydrates onto BP modified glass slides

1. For highest resolution patterning, the argon ion laser beam should be homogenized using refractive beam-shaping optics and expanded to give a uniform illumination plane. The laser power should be adjusted to give a final illumination intensity of ~14 mW/cm^2 at the substrate surface. Alternatively, a UV LED (Clearstone Technologies) can be utilized for applications where high spatial pattern resolution is not required. Adjust LED output to ~14 mW/cm^2.
2. The protein and carbohydrate patterning targets should be reconstituted or diluted according to manufacturer's instructions, aliquoted, and frozen at −80 °C until time of use. Protein solutions to be used for patterning experiments are prepared fresh for each experiment. For photopatterning applications protein solutions of concentration 5 μg/ml work well for most targets.
3. Solution phase patterning targets are introduced to the functionalized glass slides in a parallel-plate flow chamber. The chamber is separated from the glass substrate by a silicone gasket. A vacuum is applied to hold the assembly together while the bimolecular solution is pulled through the chamber via a syringe pump.
4. Once assembled, the flow chamber is positioned face down so the UV light is incident on the nonfunctionalized back of the glass slide. For spatial patterns, illumination is through chromium-coated quartz photomasks placed between the UV source and the BP-modified slides. For spatial gradients, an opaque shutter

is attached to a translational stage and the stage is moved in time so as to create a spatial gradient in exposure time across the substrate during UV exposure.

5. Following UV exposure, the slide is separated from the chamber while immersed in a buffer rinse solution to prevent drying-induced protein denaturation. Though the composition of the rinse solutions will vary based upon the biomolecule being patterned, successful past examples include: 0.5% (v/v) Tween 20 in PBS buffer for substrates presenting concanavalin A and fibronectin, 0.5 mg/ml sodium dodecyl sulfate in Dulbecco's PBS for substrates presenting mannan, and 0.5% (v/v) Tween 20 and 1% (w/v) BSA in Dulbecco's PBS with Ca^{2+} and Mg^{2+} for substrates presenting P-selectin.

6. If patterning more than one target, flow a low-pH glycine solution (pH 2.2) through the device to remove nonspecifically adsorbed biomolecules following the first exposure. Without disassembling the flow cell, then introduce PBS to first remove residual rinse solution components and then introduce the second biomolecule solution to the substrate and pattern as described earlier. Repeat the process for additional components. When finished with all patterning, disassemble the flow chamber in predetermine rinse solution.

7. Store photopatterned substrates in 1% BSA/PBS solutions at 4 °C for up to a week.

15.2 BP PATTERNING ON POROUS COLLAGEN–GLYCOSAMINOGLYCAN SCAFFOLDS

This process is based upon that described in Martin, Caliari, et al. (2011). It has been successfully applied to the spatial patterning of proteins, including concanavalin A, fibronectin, VEGF, PDGF, and BMP2. In addition to patterning on porous collagen–glycosaminoglycan (GAG) scaffolds, this method is also amenable to patterning on planar collagen–GAG membranes (Caliari, Ramirez, & Harley, 2011). A schematic illustration of the surface chemistry and photopatterning scheme used on collagen biomaterials, along with images showing representative biomolecular patterns is shown in Fig. 15.2.

15.2.1 Materials

- Freeze-dried collagen scaffolds
- BP isothiocyante (Sinsheimer, Jagodić, Polak, Hong, & Burckhalter, 1975)
- DMF (Fisher Scientific)
- Absolute ethanol (Decan Laboratories, King of Prussia, PA)
- Millipore water
- Glass microscope slides (Fisher Scientific, Philadelphia, PA)
- Rubber o-ring
- Glass cover slips (Fisher Scientific, Philadelphia, PA)
- PBS (Sigma-Aldrich)
- BSA (Sigma-Aldrich)
- Pluronic F-27 (Sigma-Aldrich)
- N,N-Diisopropylethylamine (Sigma-Aldrich)

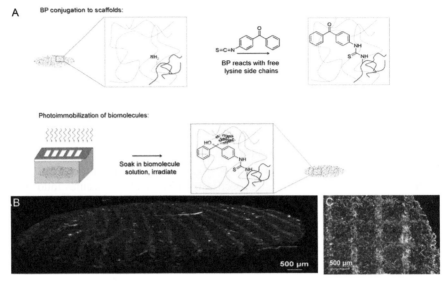

FIGURE 15.2

Photoimmobilization of biomolecules onto highly porous collagen–glycosaminoglycan scaffolds. (A) Schematic illustration showing the chemical derivatization and subsequent approach to photopatterning. (B) Photoimmobilized 100 μm stripes of biotinylated concanavalin A visualized fluorescently after postpattern labeling with Qdot 525-streptavidin. (C) Photoimmobilized 100 μm stripes of N-cadherin (horizontal, red) and fibronectin (vertical, green) as fluorescently visualized using specific antibodies. (See color plate.)

Reprinted with permission from Martin, Caliari, et al. (2011). Copyright 2011 Elsevier.

15.2.2 **Equipment**

- UV light source (Preferred: Argon ion laser, Coherent Innova 90-4)
- Refractive beam shaping optics (π-Shaper, Molecular Technologies, Berlin, Germany)
- Chromium-coated quartz photomasks
- Translational stage
- Optical power meter

15.2.3 **Method**

15.2.3.1 *BP functionalization of collagen–glycosaminoglycan scaffolds*

1. Collagen–glycosaminoglycan (CG) scaffolds are prepared via a previously reported freeze-drying technique (Harley, Leung, Silva, & Gibson, 2007; O'Brien, Harley, Yannas, & Gibson, 2004).
2. BP is immobilized when free amines on collagen are reacted with BP isothiocyante. Prepare a 20 mM solution of benzophenone-4-isothiocyanate containing 0.5 M *N,N*-diisopropylethylamine in DMF (Sinsheimer et al., 1975).

3. Add CG scaffolds to BP solution and allow to react on a shaker at room temperature in the dark for 48 h.
4. After functionalization, rinse the scaffold for 1 h at least three times in DMF and once in ethanol before final storage in water. The rinsing process is complete when the center of the scaffold is no longer yellow.

15.2.3.2 Photoattachment of biomolecules onto BP-presenting CG scaffolds

1. Adjust the UV source such that it the power is $\sim$20 mW/cm^2 at the patterning surface.
2. Prepare a solution of the biomolecule to be patterned. The concentration of each target must be optimized according to a working range of 1–100 μg/ml.
3. The scaffolds should be presoaked in a buffer solution containing the protein of interest for 1 h at 4 °C.
4. Place a rubber o-ring onto a glass slide to create a well. Carefully position the protein soaked scaffold in the middle of the rubber o-ring and pipette a 20 μl aliquot of protein solution onto the scaffold. Place a coverslip over top of the o-ring and scaffold to seal the chamber.
5. Following irradiation, immerse the scaffolds in a solution containing 0.2% pluronic F-127 in PBS for 1 h.
6. Alternatively, for a two-component patterning, wash the scaffold for 1 h in PBS following the first excitation before incubating for 1 h with the secondary component to be patterned and exposed as described earlier.
7. Postpatterning, scaffolds are blocked in a 1% (w:v) solution of BSA in PBS.

15.3 DISCUSSION

BP photopatterning is a simple, versatile surface patterning technique. By modulating UV exposure across BP functionalized substrates, spatial selectivity and controllable protein site density can be achieved. Slight modifications to the protocol may be required based upon the particular biomolecule of interest to be patterned. Common modifications include a change in biomolecule concentration and/or UV exposure time. This methodology is molecularly general and, due to a nonreliance on physical contact for patterning, as is the case with stamp-based patterning approaches, amenable to nonplanar substrates, including topographically complex substrates. The approach has also been demonstrated on corrugated glass slides, in addition to highly porous and nonporous collagen biomaterials.

Despite the enabling capabilities of this biomolecular patterning approach, an obvious concern lies in the ultimate bioactivity of the resulting patterns given the exposure to UV light. However, a several published and unpublished studies have confirmed that many different biomolecules still retain their native biological effect after photochemical patterning. Two particular investigations are presented below as examples.

15.3.1 Example: leukocyte rolling and adhesion on photochemically patterned substrates

Irregular inflammatory and immune responses can result in a variety of diseases including rheumatoid arthritis, asthma, psoriasis, Crohn's disease, thrombotic disorders, and autoimmune disease (Schmid-Schonbein, 2006). BP photoimmobilization of biomolecular gradients and patterns were used to conduct a high-throughput investigation of the first step of the inflammatory response, leukocyte recruitment. A family of molecules known as selectins have been identified as key players in the initial flow-dependent tethering and rolling of leukocytes on vascular surfaces (Bevilacqua et al., 1991; Lasky, 1992; McEver, 1994, 2002). HL-60 promyelocytes and Jurkat T lymphocytes were introduced to substrates presenting gradients of either P-selectin or PSGL-1 under condition of physiological shear stress. Cell behavior was recorded at each combination of site density and shear stress. It was shown that HL-60 cells tether and roll on one-component P-selectin gradient substrates over a wide range of site densities and wall shear stresses. Jurkat cells also tethered to and rolled on immobilized P-selectin, as well as PSGL-1, over a range of site densities and shear stresses. This proof-of-principle study demonstrates the efficacy of BP patterning for the presentation of biologically active ligands at controllably varied site densities on a single substrate, allowing for large amounts of data to be acquired with minimal cost of time, materials, and effort.

15.3.2 Example: enhanced cell attachment to photochemically patterned collagen scaffolds

BP photopatterning was successfully demonstrated on three-dimensional collagen–GAG scaffolds. No significant changes to the bulk mechanical properties of BP-modified scaffolds in response to the organic modification or exposure to UV light were observed. The efficacy of CG scaffolds as culture substrates was assessed by monitoring the proliferation of MC3T3-E1 preosteoblasts seeded onto scaffolds both with and without BP. It was shown that the addition of BP does not have a negative impact on scaffold bioactivity. Fibronectin was photoimmobilized on CB–BP scaffolds to demonstrate the ability to elicit a specific biological response with photoimmobilized fibronectin promoting greater MC3T3-E1 cell attachment compared to scaffolds containing physiosorbed fibronectin or scaffolds that were not exposed to fibronectin. These results demonstrate that BP photoimmobilization is a viable method to introduce instructive biomolecular cues within collagen–GAG biomaterials for future studies in the areas of regenerative medicine and tissue engineering.

GENERAL CONCLUSIONS

BP-mediated photochemical patterning offers a direct approach to generate complex, multicomponent patterns or gradients on a range of substrate materials. Since this immobilization approach only requires the presence of a C–H bond, it is biomolecularly general and has broad applicability to a range of classes of biomolecules. Photopatterned glass and collagen biomaterials have been demonstrated as potential model substrate for subsequent studies of cellular behavior. The BP moiety can be conveniently tethered to many substrate materials using straightforward conjugation chemistries and UV exposure to radiation in the ~350 nm wavelength range does not inactivate or otherwise damage biomolecules immobilized on the substrate surface. While some target biomolecules have been found to be more easily patterned using this approach than others, the general capability to create multicomponent biomolecular patterns and gradient with control over biomolecular deposition density should prove to be quite enabling for a myriad of studies investigating the biology–materials interface.

Acknowledgments

We acknowledge support for our efforts to develop photolithographic biomolecular patterning schemes from the National Science Foundation, through award DMR-1105300, as well as from the Roy J. Carver Charitable Trust, the Camille and Henry Dreyfus Foundation, and 3M.

References

Bevilacqua, M., Butcher, E., Furie, B., Furie, B., Gallatin, M., Gimbrone, M., et al. (1991). Selectins: A family of adhesion receptors. *Cell, 67*(2), 233.

Caliari, S. R., Ramirez, M. A., & Harley, B. A. C. (2011). The development of collagen–GAG scaffold-membrane composites for tendon tissue engineering. *Biomaterials, 32*(34), 8990–8998.

Harder, P., Grunze, M., Dahint, R., Whitesides, G. M., & Laibinis, P. E. (1998). Molecular conformation in oligo(ethylene glycol)-terminated self-assembled monolayers on gold and silver surfaces determines their ability to resist protein adsorption. *Journal of Physical Chemistry B, 102*(2), 426–436.

Harley, B. A., Leung, J. H., Silva, E C C M, & Gibson, L. J. (2007). Mechanical characterization of collagen–glycosaminoglycan scaffolds. *Acta Biomaterialia, 3*(4), 463–474.

Herman, C. T., Potts, G. K., Michael, M. C., Tolan, N. V., & Bailey, R. C. (2011). Probing dynamic cell-substrate interactions using photochemically generated surface-immobilized gradients: Application to selectin-mediated leukocyte rolling. *Integrative Biology, 3*(7), 779–791.

Lasky, L. A. (1992). Selectins: Interpreters of cell-specific carbohydrate information during inflammation. *Science, 258*(5084), 964–969.

Lawrence, M. B., & Springer, T. A. (1991). Leukocytes roll on a selectin at physiologic flow rates: Distinction from and prerequisite for adhesion through integrins. *Cell (Cambridge, MA), 65*(5), 859–873.

Martin, T. A., Caliari, S. R., Williford, P. D., Harley, B. A., & Bailey, R. C. (2011). The generation of biomolecular patterns in highly porous collagen–GAG scaffolds using direct photolithography. *Biomaterials, 32*(16), 3949–3957.

Martin, T. A., Herman, C. T., Limpoco, F. T., Michael, M. C., Potts, G. K., & Bailey, R. C. (2011). Quantitative photochemical immobilization of biomolecules on planar and corrugated substrates: A versatile strategy for creating functional biointerfaces. *ACS Applied Materials and Interfaces, 3*(9), 3762–3771.

McEver, R. P. (1994). Selectins. *Current Opinion in Immunology, 6*(1), 75–84.

McEver, R. P. (2002). Selectins: Lectins that initiate cell adhesion under flow. *Current Opinion in Cell Biology, 14*(5), 581–586.

O'Brien, F. J., Harley, B. A., Yannas, I. V., & Gibson, L. (2004). Influence of freezing rate on pore structure in freeze-dried collagen–GAG scaffolds. *Biomaterials, 25*(6), 1077–1086.

Raghavan, S., Desai, R. A., Kwon, Y., Mrksich, M., & Chen, C. S. (2010). Micropatterned dynamically adhesive substrates for cell migration. *Langmuir, 26*(22), 17733–17738.

Schmid-Schonbein, G. W. (2006). Analysis of inflammation. *Annual Review of Biomedical Engineering, 8*, 93–151.

Sinsheimer, J. E., Jagodić, V., Polak, L., Hong, D. D., & Burckhalter, J. H. (1975). Polycyclic aromatic isothiocyanate compounds as fluorescent labeling reagents. *Journal of Pharmaceutical Sciences, 64*(6), 925–930.

Tibbitt, M. W., & Anseth, K. S. (2012). Dynamic microenvironments: The fourth dimension. *Science Translational Medicine, 4*(160), 160ps124.

Toh, C. R., Fraterman, T. A., Walker, D. A., & Bailey, R. C. (2009). Direct biophotolithographic method for generating substrates with multiple overlapping biomolecular patterns and gradients. *Langmuir, 25*(16), 8894–8898.

Index

Note: Page numbers followed by *f* indicate figures and *t* indicate tables.

243

Volumes in Series

Founding Series Editor
DAVID M. PRESCOTT

Volume 1 (1964)
Methods in Cell Physiology
Edited by David M. Prescott

Volume 2 (1966)
Methods in Cell Physiology
Edited by David M. Prescott

Volume 3 (1968)
Methods in Cell Physiology
Edited by David M. Prescott

Volume 4 (1970)
Methods in Cell Physiology
Edited by David M. Prescott

Volume 5 (1972)
Methods in Cell Physiology
Edited by David M. Prescott

Volume 6 (1973)
Methods in Cell Physiology
Edited by David M. Prescott

Volume 7 (1973)
Methods in Cell Biology
Edited by David M. Prescott

Volume 8 (1974)
Methods in Cell Biology
Edited by David M. Prescott

Volume 9 (1975)
Methods in Cell Biology
Edited by David M. Prescott

Volume 10 (1975)
Methods in Cell Biology
Edited by David M. Prescott

Advisory Board Chairman
KEITH R. PORTER

Series Editor
LESLIE WILSON

Volume 30 (1989)
Fluorescence Microscopy of Living Cells in Culture, Part B: Quantitative Fluorescence Microscopy—Imaging and Spectroscopy
Edited by D. Lansing Taylor and Yu-Li Wang

Volume 31 (1989)
Vesicular Transport, Part A
Edited by Alan M. Tartakoff

Volume 32 (1989)
Vesicular Transport, Part B
Edited by Alan M. Tartakoff

Volume 33 (1990)
Flow Cytometry
Edited by Zbigniew Darzynkiewicz and Harry A. Crissman

Volume 34 (1991)
Vectorial Transport of Proteins into and across Membranes
Edited by Alan M. Tartakoff

Selected from Volumes 31, 32, and 34 (1991)
Laboratory Methods for Vesicular and Vectorial Transport
Edited by Alan M. Tartakoff

Volume 35 (1991)
Functional Organization of the Nucleus: A Laboratory Guide
Edited by Barbara A. Hamkalo and Sarah C. R. Elgin

Volume 36 (1991)
***Xenopus laevis:* Practical Uses in Cell and Molecular Biology**
Edited by Brian K. Kay and H. Benjamin Peng

Series Editors
LESLIE WILSON AND PAUL MATSUDAIRA

Volume 37 (1993)
Antibodies in Cell Biology
Edited by David J. Asai

Volume 38 (1993)
Cell Biological Applications of Confocal Microscopy
Edited by Brian Matsumoto

Volume 39 (1993)
Motility Assays for Motor Proteins
Edited by Jonathan M. Scholey

Volume 40 (1994)
A Practical Guide to the Study of Calcium in Living Cells
Edited by Richard Nuccitelli

Volume 41 (1994)
Flow Cytometry, Second Edition, Part A
Edited by Zbigniew Darzynkiewicz, J. Paul Robinson, and Harry A. Crissman

Volume 42 (1994)
Flow Cytometry, Second Edition, Part B
Edited by Zbigniew Darzynkiewicz, J. Paul Robinson, and Harry A. Crissman

Volume 43 (1994)
Protein Expression in Animal Cells
Edited by Michael G. Roth

Volume 44 (1994)
Drosophila melanogaster: **Practical Uses in Cell and Molecular Biology**
Edited by Lawrence S. B. Goldstein and Eric A. Fyrberg

Volume 45 (1994)
Microbes as Tools for Cell Biology
Edited by David G. Russell

Volume 46 (1995)
Cell Death
Edited by Lawrence M. Schwartz and Barbara A. Osborne

Volume 47 (1995)
Cilia and Flagella
Edited by William Dentler and George Witman

Volume 48 (1995)
Caenorhabditis elegans: **Modern Biological Analysis of an Organism**
Edited by Henry F. Epstein and Diane C. Shakes

Volume 49 (1995)
Methods in Plant Cell Biology, Part A
Edited by David W. Galbraith, Hans J. Bohnert, and Don P. Bourque

Volume 62 (1999)
Tetrahymena thermophila
Edited by David J. Asai and James D. Forney

Volume 63 (2000)
Cytometry, Third Edition, Part A
Edited by Zbigniew Darzynkiewicz, J. Paul Robinson,
   and Harry Crissman

Volume 64 (2000)
Cytometry, Third Edition, Part B
Edited by Zbigniew Darzynkiewicz, J. Paul Robinson,
   and Harry Crissman

Volume 65 (2001)
Mitochondria
Edited by Liza A. Pon and Eric A. Schon

Volume 66 (2001)
Apoptosis
Edited by Lawrence M. Schwartz and Jonathan D. Ashwell

Volume 67 (2001)
Centrosomes and Spindle Pole Bodies
Edited by Robert E. Palazzo and Trisha N. Davis

Volume 68 (2002)
Atomic Force Microscopy in Cell Biology
Edited by Bhanu P. Jena and J. K. Heinrich Hörber

Volume 69 (2002)
Methods in Cell–Matrix Adhesion
Edited by Josephine C. Adams

Volume 70 (2002)
Cell Biological Applications of Confocal Microscopy
Edited by Brian Matsumoto

Volume 71 (2003)
Neurons: Methods and Applications for Cell Biologist
Edited by Peter J. Hollenbeck and James R. Bamburg

Volume 72 (2003)
Digital Microscopy: A Second Edition of Video Microscopy
Edited by Greenfield Sluder and David E. Wolf

Volume 73 (2003)
Cumulative Index

Volume 74 (2004)
Development of Sea Urchins, Ascidians, and Other
Invertebrate Deuterostomes: Experimental Approaches
Edited by Charles A. Ettensohn, Gary M. Wessel,
and Gregory A. Wray

Volume 75 (2004)
Cytometry, 4th Edition: New Developments
Edited by Zbigniew Darzynkiewicz, Mario Roederer,
and Hans Tanke

Volume 76 (2004)
The Zebrafish: Cellular and Developmental Biology
Edited by H. William Detrich, III, Monte Westerfield,
and Leonard I. Zon

Volume 77 (2004)
The Zebrafish: Genetics, Genomics, and Informatics
Edited by William H. Detrich, III, Monte Westerfield,
and Leonard I. Zon

Volume 78 (2004)
Intermediate Filament Cytoskeleton
Edited by M. Bishr Omary and Pierre A. Coulombe

Volume 79 (2007)
Cellular Electron Microscopy
Edited by J. Richard McIntosh

Volume 80 (2007)
Mitochondria, 2nd Edition
Edited by Liza A. Pon and Eric A. Schon

Volume 81 (2007)
Digital Microscopy, 3rd Edition
Edited by Greenfield Sluder and David E. Wolf

Volume 82 (2007)
Laser Manipulation of Cells and Tissues
Edited by Michael W. Berns and Karl Otto Greulich

Volume 83 (2007)
Cell Mechanics
Edited by Yu-Li Wang and Dennis E. Discher

Volume 84 (2007)
Biophysical Tools for Biologists, Volume One: In Vitro
Techniques
Edited by John J. Correia and H. William Detrich, III

Volume 97 (2010)
Microtubules: In Vivo
Edited by Lynne Cassimeris and Phong Tran

Volume 98 (2010)
Nuclear Mechanics & Genome Regulation
Edited by G.V. Shivashankar

Volume 99 (2010)
Calcium in Living Cells
Edited by Michael Whitaker

Volume 100 (2010)
The Zebrafish: Cellular and Developmental Biology,
   Part A
Edited by: H. William Detrich III, Monte Westerfield
   and Leonard I. Zon

Volume 101 (2011)
The Zebrafish: Cellular and Developmental Biology,
   Part B
Edited by: H. William Detrich III, Monte Westerfield and
   Leonard I. Zon

Volume 102 (2011)
Recent Advances in Cytometry, Part A: Instrumentation,
   Methods
Edited by Zbigniew Darzynkiewicz, Elena Holden, Alberto Orfao,
   William Telford and Donald Wlodkowic

Volume 103 (2011)
Recent Advances in Cytometry, Part B: Advances in
   Applications
Edited by Zbigniew Darzynkiewicz, Elena Holden, Alberto Orfao,
   Alberto Orfao and Donald Wlodkowic

Volume 104 (2011)
The Zebrafish: Genetics, Genomics and Informatics
   3rd Edition
Edited by H. William Detrich III, Monte Westerfield,
   and Leonard I. Zon

Volume 105 (2011)
The Zebrafish: Disease Models and Chemical Screens
   3rd Edition
Edited by H. William Detrich III, Monte Westerfield,
   and Leonard I. Zon

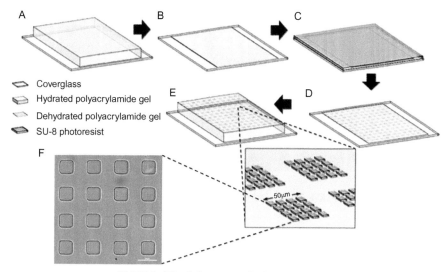

Coverglass
Hydrated polyacrylamide gel
Dehydrated polyacrylamide gel
SU-8 photoresist

50μm

PLATE 1 (Fig. 1.1 on page 6 of this volume).

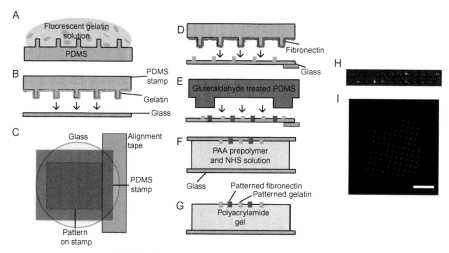

A Fluorescent gelatin solution
PDMS

B PDMS stamp
Gelatin
Glass

C Glass
Alignment tape
PDMS stamp
Pattern on stamp

D Fibronectin
Glass

E Gluteraldehyde treated PDMS

F PAA prepolymer and NHS solution
Glass

G Patterned fibronectin
Patterned gelatin
Polyacrylamide gel

H

I

PLATE 2 (Fig. 2.1 on page 22 of this volume).

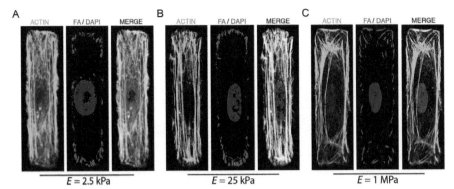

PLATE 3 (Fig. 3.4 on page 45 of this volume).

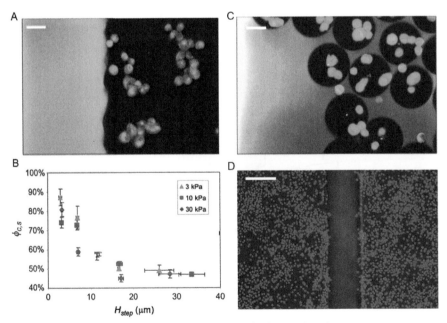

PLATE 4 (Fig. 4.2 on page 58 of this volume).

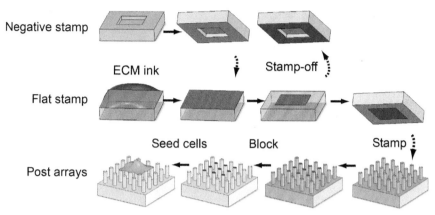

Negative stamp

ECM ink   Stamp-off

Flat stamp

Seed cells   Block   Stamp

Post arrays

PLATE 5 (Fig. 5.1 on page 62 of this volume).

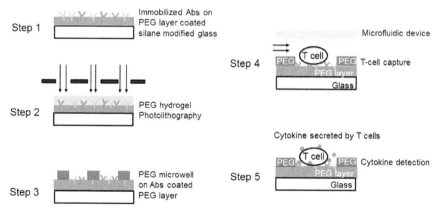

Step 1 — Immobilized Abs on PEG layer coated silane modified glass

Step 2 — PEG hydrogel Photolithography

Step 3 — PEG microwell on Abs coated PEG layer

Step 4 — Microfluidic device — T cell — PEG — T-cell capture — PEG layer — Glass

Cytokine secreted by T cells

Step 5 — T cell — PEG — Cytokine detection — PEG layer — Glass

PLATE 6 (Fig. 6.1 on page 77 of this volume).

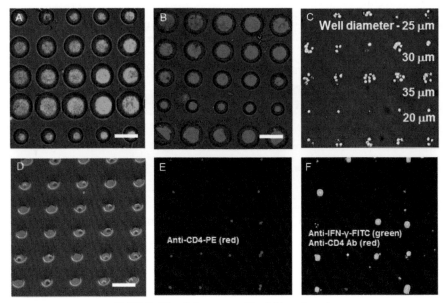

PLATE 7 (Fig. 6.2 on page 81 of this volume).

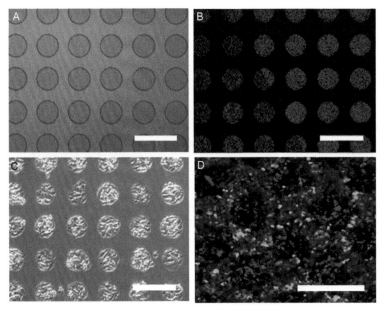

PLATE 8 (Fig. 6.4 on page 85 of this volume).

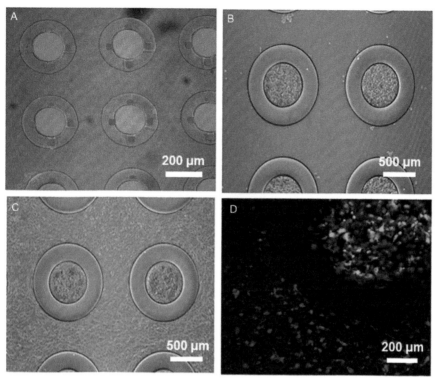

PLATE 9 (Fig. 6.5 on page 86 of this volume).

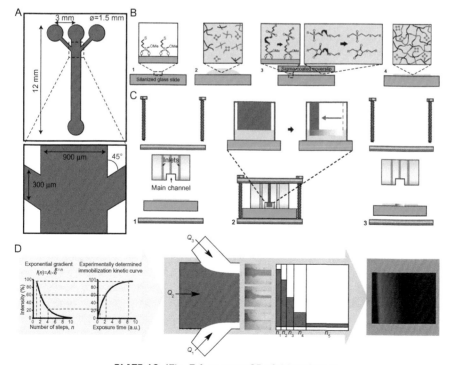

PLATE 10 (Fig. 7.1 on page 95 of this volume).

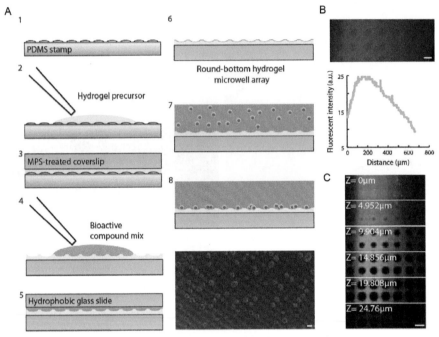

PLATE 11 (Fig. 7.2 on page 96 of this volume).

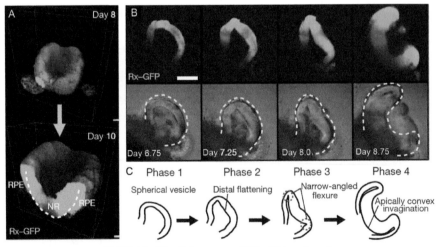

PLATE 12 (Fig. 9.1 on page 123 of this volume).

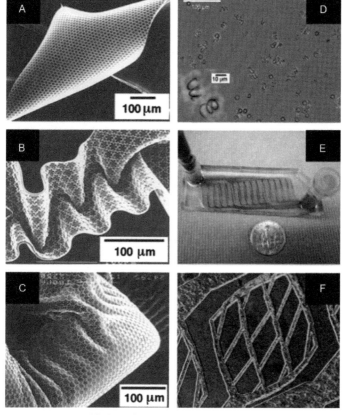

PLATE 13 (Fig. 9.3 on page 128 of this volume).

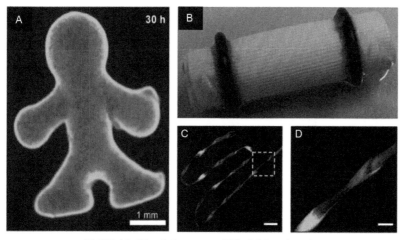

PLATE 14 (Fig. 9.4 on page 129 of this volume).

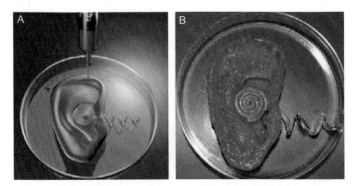

PLATE 15 (Fig. 9.5 on page 130 of this volume).

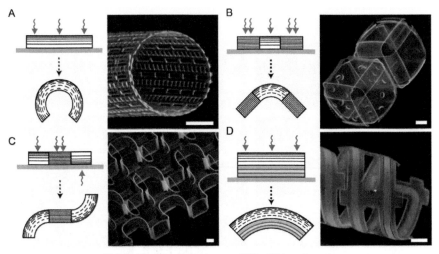

PLATE 16 (Fig. 9.6 on page 132 of this volume).

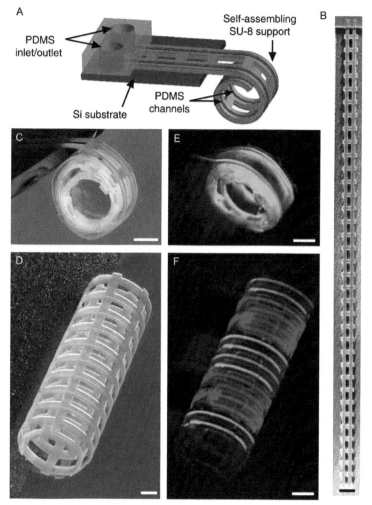

PLATE 17 (Fig. 9.7 on page 133 of this volume).

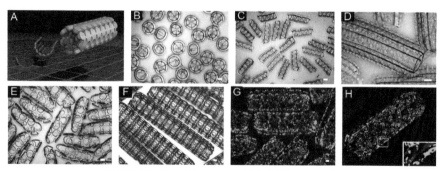

PLATE 18 (Fig. 9.8 on page 134 of this volume).

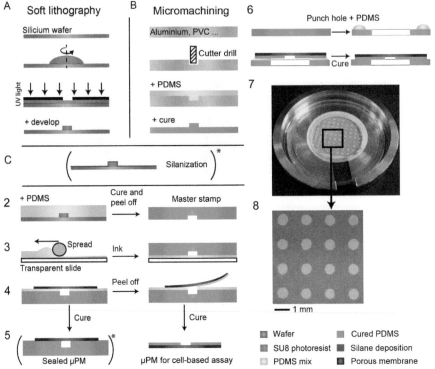

PLATE 19 (Fig. 11.1 on page 160 of this volume).

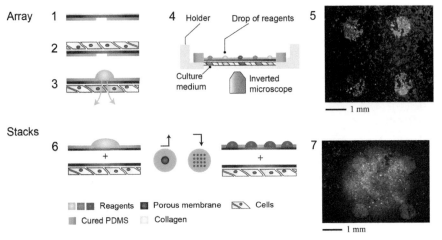

PLATE 20 (Fig. 11.2 on page 168 of this volume).

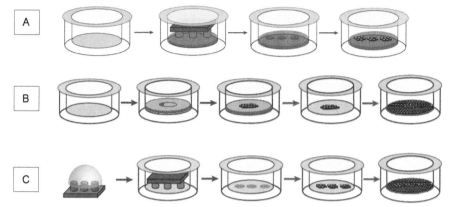

PLATE 21 (Fig. 12.1 on page 174 of this volume).

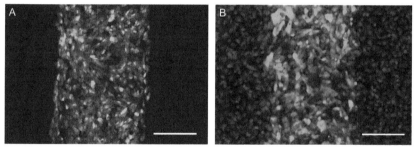

PLATE 22 (Fig. 12.2 on page 181 of this volume).

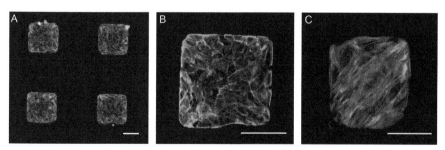

PLATE 23 (Fig. 12.3 on page 181 of this volume).

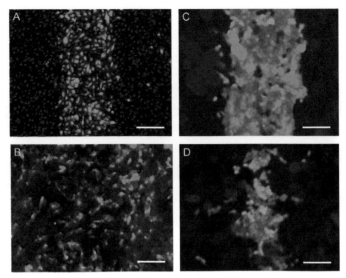

PLATE 24 (Fig. 12.5 on page 184 of this volume).

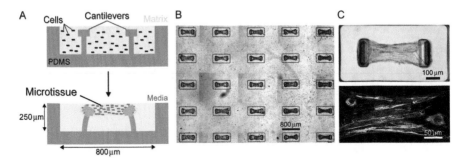

PLATE 25 (Fig. 13.1 on page 193 of this volume).

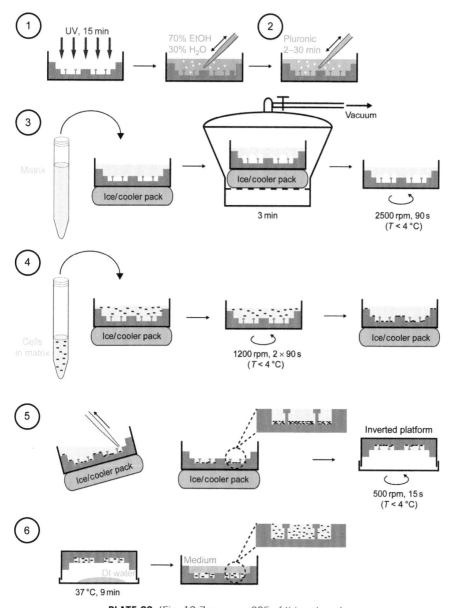

PLATE 26 (Fig. 13.7 on page 205 of this volume).

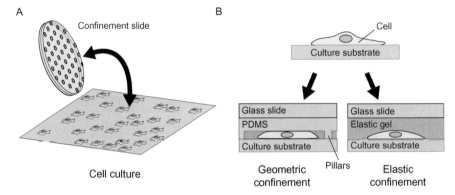

PLATE 27 (Fig. 14.1 on page 215 of this volume).

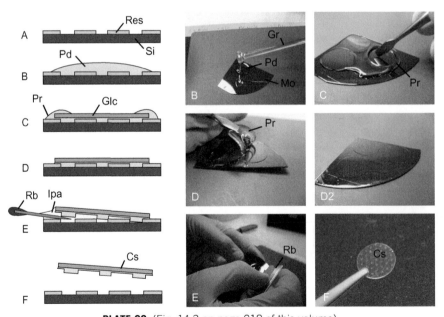

PLATE 28 (Fig. 14.3 on page 218 of this volume).

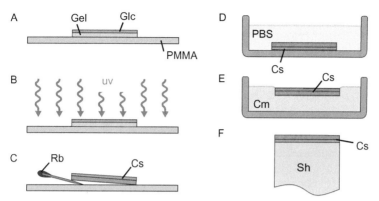

PLATE 29 (Fig. 14.4 on page 221 of this volume).

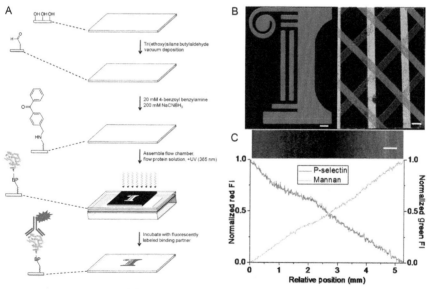

PLATE 30 (Fig. 15.1 on page 234 of this volume).

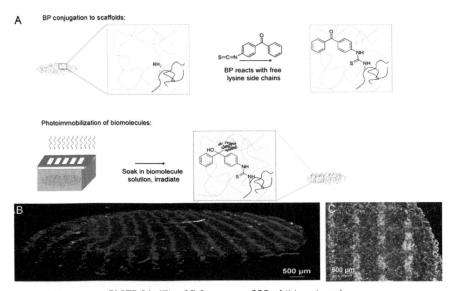

PLATE 31 (Fig. 15.2 on page 238 of this volume).